THE ADAPTIVE
GEOMETRY OF
Trees

MONOGRAPHS IN POPULATION BIOLOGY

EDITED BY ROBERT M. MAY

1. The Theory of Island Biogeography, by Robert H. Mac-Arthur and Edward O. Wilson
2. Evolution in Changing Environments: Some Theoretical Explorations, by Richard Levins
3. Adaptive Geometry of Trees, by Henry S. Horn
4. Theoretical Aspects of Population Genetics, by Motoo Kimura and Tomoko Ohta
5. Populations in a Seasonal Environment, by Stephen D. Fretwell
6. Stability and Complexity in Model Ecosystems, by Robert M. May
7. Competition and the Structure of Bird Communities, by Martin L. Cody
8. Sex and Evolution, by George C. Williams
9. Group Selection in Predator-Prey Communities, by Michael E. Gilpin
10. Geographic Variation, Speciation, and Clines, by John A. Endler
11. Food Webs and Niche Space, by Joel E. Cohen
12. Caste and Ecology in the Social Insects, by George F. Oster and Edward O. Wilson
13. The Dynamics of Arthropod Predator-Prey Systems, by Michael P. Hassell
14. Some Adaptations of Marsh-Nesting Blackbirds, by Gordon H. Orians
15. Evolutionary Biology of Parasites, by Peter W. Price
16. Cultural Transmission and Evolution: A Quantitative Approach, by L. L. Cavalli-Sforza and M. W. Feldman
17. Resource Competition and Community Structure, by David Tilman
18. The Theory of Sex Allocation, by Eric L. Charnov
19. Mate Choice in Plants: Tactics, Mechanisms, and Consequences, by Mary F. Willson and Nancy Burley

THE ADAPTIVE
GEOMETRY OF
Trees

BY HENRY S. HORN

PRINCETON, NEW JERSEY
PRINCETON UNIVERSITY PRESS

Contents

Preface   vii

Preface to the Second Printing   ix

Symbols Used   xii

1. Introduction   3

2. Measurement of Light Intensity   8

3. Analysis of a Forest Succession   19

4. Theoretical Strategies of Leaf Distribution   45

5. Photosynthetic Response of the Strategies   64

6. Measurement of Actual Strategies   89

7. Speculations on the Shapes of Tree Crowns   104

8. On the Relation between Theory and Reality   118

Nomenclature   131

Bibliography   135

Index   141

Preface

I started this project almost by accident three years ago. Robert MacArthur and I had been trying to find mechanical ways to measure foliage profiles, which he had found to be accurate predictors of the variety of birds in a habitat. One of our machines was an adjustable light meter that I shall describe in Chapter 2. While testing it in the woods on a Sunday afternoon, I recorded the meter readings above various saplings, and I was struck by a general pattern. Trees that typically invade old fields were never found under a canopy. The familiar species of young forests were found beneath openings in the canopy, and denizens of virgin forests were found in deep shade. An accumulation of many more measurements eventually gave strong support to the foresters' axiom that species which tolerate a canopy become progressively more dominant as succession proceeds — a result that was more gratifying than surprising. However, the analysis raised new questions and cast new light on some old ones. I subsequently invented a theory to explore these questions. The theory led to new kinds of measurements, which led to changes in the theory, and so on. The result to date is this book, in which I simultaneously commit the sins of vitalism and mechanism. Indeed, thinking of trees as crafty green strategists has given me many new insights, and simplified assumptions have allowed me to test these insights. I have organized my ideas around the traditional problem of plant succession, but my basic interest is the prediction and testing of patterns of adaptive strategies among trees. I sincerely hope that my ideas will not be viewed as a contribution to past and current controversies about the appropriateness of successional terminologies.

Rather, I am much more interested in the way that nature shapes theory than in the constraints that theory imposes on nature. The greatest homage that can be paid to an empirical theory is the constructive criticism that makes it obsolete at an early age.

Throughout this study I have had continual discussions with Robert MacArthur and Egbert Leigh. I am indebted to them for elegant improvements in the theory and for critical empirical observations. I have also had very helpful comments and observations from John Bonner, Gordon Orians, Daniel Janzen, Christopher Smith, Woodruff Benson, and J. Merritt Emlen.

For permission to work on their land, or for tolerating my presence without permission, I am grateful to those in charge of the Institute Woods, Herrontown Woods County Park, and the Stony Ford Field Station at Princeton, N.J.; Mammoth Cave National Park, Ky.; Calaveras County Big Tree Park, Calif.; and Finca las Cruces, San Vito de Java, Puntarenas Prov., Costa Rica. For help in the field and in the lab, I am indebted to my wife, Elizabeth, to Thomas Gibson, to Robert Millaway, and to several students in the August 1969 O.T.S. Advanced Population Biology group.

The funds for this study were scrounged from many sources. My wife allowed me to use part of the family budget that she earned. The Eugene Higgens Trust Fund of the Princeton Department of Biology provided pin money. I had free use of electronic gadgets provided and maintained by the Whitehall Foundation. The Organization for Tropical Studies paid my way to and in Costa Rica in return for a series of lectures on theoretical forestry. Although they don't know it yet, the Princeton University Computer Center provided free computer time, charged to a grant from the National Science Foundation.

For reading the entire manuscript and tendering valuable criticisms and suggestions, I am exceedingly thankful to Robert MacArthur, Egbert Leigh, Christopher Smith, Howard Howland, Elizabeth Horn, John Bonner, Gordon Orians, and David Horn. Thomas Frazzetta, Richard Horn, Daniel Janzen, Otto Solbrig, and Carl Gans read and commented on parts of the manuscript.

The book is dedicated to my father, Henry E. Horn, who introduced me to forests when I was four years old.

Preface to the Second Printing (1976)

Since 1971 there have been several developments in the literature and in my own thinking that would make this book different were I to write it now. The outline, arguments, and results would be the same, but I would change some of the following details.

Chapters 4-6 give the impression that the technical development of the theory rests heavily on the assumption that leaves are randomly distributed disks. However, Equations 5.8, 5.10, and 5.13 could be derived with even simpler notation from an arbitrary index of leaf area per unit of ground area, with leaves distributed into independent layers, within each of which the leaves are evenly dispersed. The models are not restricted to a peculiar leaf shape or to a particular vertical distribution of leaves.

I have recently described the difference between mono-layered and multilayered trees more graphically (1975. *Scientific American* 232 (5): 90-98). The essential difference is less between different vertical distributions of branches, than between a tree with its leaves in a continuous shell about the peripheral twigs and a tree with leaves scattered loosely throughout its volume (cf. page 63). In the same article, I address the question of why individual trees tend to be either monolayered or multilayered, with only limited flexibility between these extremes. Whitney (1976. *Bull. Torrey Bot. Club*, in press) has related leaf distribution to branching structure. I have expanded the discussion in Chapter 8 of patterns of productivity, stability, and diversity in succession (1974. *Ann. Rev. Ecol. Syst.* 5: 25-37).

In Chapters 2 and 6, I leave the impression that special equipment is needed to measure the proportion of light that directly penetrates a branch or a tree. However, for measuring "layers" in a tree, surprising accuracy can be attained by rolling up a piece of paper into a tube, sighting the appropriate branch or tree, adjusting the width of the tube or its distance from the eye until only the section to be measured is seen, and guessing the proportion of sky that is unobscured. Accuracy in the range between 0.1 and 0.9 comes quickly with experience. Beyond this range even a light meter is often inaccurate. Accuracy is not critical however. Similar errors affect the measurement of both branch and tree. When several Princeton undergraduates have measured the same tree, their raw data seem appallingly disparate, but the ratio of logarithms of each individual's measurements, that is the number of "layers," shows little variation between observers. The errors that persist involve very sparsely foliated branches on very shady trees. These are multilayers, for which compulsive accuracy is fatuous because the photosynthetic difference between 2 and 5 layers is about the same as that between 5 layers and an infinite number (page 78).

Monsi, Uchijima, and Oikawa (1973. *Ann. Rev. Ecol. Syst.* 4: 301-327) review more accurate and perforce less general studies of the effect of canopy structure on photosynthesis. The cyclic persistence of Beech and Sugar Maple, inferred on pages 35-37, has been explored by Forcier (1975. *Science* 189: 808-810), who interprets their presence with Yellow Birch (*Betula alleghaniensis*) in central New Hampshire as a "climax microsuccession." The inferred allelo-pathic effect of Sassafras (pages 40-41) has been documented and explored by Gant and Clebsch (1975. *Ecology* 56: 604-615). Anderson and Miller (1974. *J. Appl. Ecol.* 11: 691-697) discuss the penumbral effect of leaves (pages 46-47) and sunflecks in the context of some striking pictures. Dahl (1973. *Marine Biol.* 23: 239-249) has invented geometric models to describe corals in a way that is more appropriate than the suggestion on page 129.

An approach analogous to mine can be applied to the stratification of trees themselves in storied forests, as Smith (1973. *Am. Naturalist* 107: 671-683) has done verbally, and J. W. Terborgh has done analytically in a lecture that he promises to publish soon. Terborgh's ideas will be reviewed in a forthcoming book by F. Hallé, R.A.A. Oldeman, and P. B. Tomlinson (in preparation. *Tree Architecture in Tropical Forest Ecosystems.* Springer Verlag, New York & Heidelberg). This book will also discuss developmental patterns that may set limitations on the shapes and leaf distributions of trees of a given species.

Symbols Used

This is not a complete glossary. It includes only symbols that are used in several places. The definitions of all other symbols may be found within a few paragraphs of where they are used. The dimensions of each term follow in parentheses. Throughout the discussions, compatible units are used for all terms. Moreover ρ is always multiplied by r^2, P_{max} is set equal to 1.0, and R is expressed as a proportion of P_{max}. These conventions make the following terms either proportions or counting-numbers; hence they are all dimensionless.

C The compensation point, the light intensity at which photosynthesis just balances respiration (proportion of full sunlight).

ϵ The base of natural logarithms and exponentials (2.718).

k A binding constant that measures the effectiveness of a leaf in getting and processing photons (units of light intensity). k is further defined in the discussion preceding Equation 5.3.

L_0 Incident light intensity (proportion of full sunlight).

log   All logarithms used in this book are natural logarithms, to the base ϵ or 2.303 times logarithm to the base 10. Exponentials and natural logarithms figure prominently and conveniently in Poisson random distributions, limits of integral powers, and solutions of integrals with the integrand in the denominator.

n Number of equally dense layers of leaves in a tree (number of layers). Its theoretical importance begins in Expression 5.1 and continues through Equation 5.13. A practical measure is defined by

Equation 6.1. The various expressions for rates of photosynthesis for trees with varying numbers of layers are discussed in the paragraph following Equation 5.13.

P Net rate of photosynthesis for a tree (net units of photosynthate produced or net CO_2 uptake/ ground area/time).

P_{max} Total rate of photosynthesis for a leaf fully lit at the optimum light intensity (total units of photosynthate produced or total CO_2 uptake/leaf area/ time).

R Rate of respiration for a leaf (units of photosynthate used or CO_2 evolved/leaf area/time).

r Radius of a circular leaf (length).

π Circumference of a circle of unit diameter (3.142).

ρ Average number of leaves, totaled for all layers, above a unit of ground area (number of leaves/ area).

∞ Infinity.

THE ADAPTIVE
GEOMETRY OF
Trees

Trees growing in a sunny or windy position are more branched shorter and less straight. And in general mountain trees have more knots than those of the plain, and those that grow in dry spots than those that grow in marshes.

Yew *pados* and joint-firs rejoice exceedingly in shade. Others, one may say in general, prefer a sunny position. However this too depends partly on the soil appropriate to each tree.

And in general those that have flat leaves have them in a regular series, as myrtle, while in other instances the leaves are in no particular order, but set at random, as in most other plants.

Trees may destroy one another, by robbing them of nourishment and hindering them in other ways. Again some things, though they do not cause death, enfeeble the tree as to the production of flavours and scents.

Theophrastus (b. 370 B.C.)

Quoted by permission of the publisher from *Enquiry into Plants,* translated by A. F. Hort, Loeb Classical Library, Harvard University Press, Cambridge, Massachusetts, 1916, pp. 291, 57, 289, 75, and 411–413.

CHAPTER ONE

Introduction

When a field is left fallow for a long time in a region of moderate climate, it returns to forest through a series of gradual changes that we call succession. Succession in temperate forests starts with the invasion of a field by pioneer species that are often characteristic of the region: for example, Eastern Redcedar[1] or Gray Birch in the northeastern United States, but pines in the southeast. There follows a highly variable series of changes in the specific composition of the stand, but eventually a stage is reached when the changes are imperceptibly slow. This stage is called the climax and is defined as that stage when no significant change occurs during the lifetimes of several research grants. Again the climax appears to be characteristic of the region: for example, in New Jersey, American Beech in moist areas and oak forests in dry areas, but in Maine and southeastern Canada, spruce forests.

Since certain species occupy characteristic successional positions in a given locality, each species is presumed to be best adapted to the environment at a particular stage. Hence to describe succession and to understand why it occurs, we must answer three questions: How does the environment change as succession proceeds? How are different species adapted to each environment in the sequence? And how does each species influence, or perhaps effect, the changes in the environment?

The most dramatic environmental change from field to

[1] Throughout the text I shall use vernacular names, which are capitalized only when they refer to a particular species. A list of the corresponding Latin names follows Chapter 8.

forest is the decrease in light intensity near the ground. Light is cut off by progressively more plants of greater height, and only a tiny fraction of the intercepted light is converted to stored energy; the rest must ultimately be dissipated as heat. Thus the distribution of heat varies as succession proceeds. The distributions of heat and of leaves affect evaporation and water balance. Water balance, in turn, affects the supply of nutrients from the soil.

Light, heat, water, and nutrient distributions all change as succession proceeds. If we describe the changes in these four factors and discover how plants are adapted to them, we can begin to understand succession. There are at least two ways to find how these factors affect the course of succession. One way is to make arbitrary measurements in factorial fashion: that is, in areas where all possible values of all factors are represented in all combinations. The data are then analyzed for the significance of each factor, using a statistical model that assumes that the factors have linear effects that combine additively. This is a powerful technique for weeding out irrelevant factors. However, if we already know that all the factors are important and if nature does not provide factorial data, or if the effects of any factors are not both linear and additive, then the technique can give answers that range from misleading to wrong.

A second technique involves biological judgment, and currently masquerades as "systems simulation" or as "component analysis," depending on whether its practitioners use digital computers or desk calculators. The factors are first listed in order of the generality of their guessed effects: light, heat, water, nutrients; the effect of the most general factor is then modeled. An ideal model should specify exactly how to make relevant measurements and exactly what mathematical form the effect of that factor should take. When the model is tested, departures from its predictions may help us to guess how the next factor in the list should be added to the model. After this process has been repeated several times, we have an exact model of the effects of all factors and their interactions.

I have applied the latter approach to explore forests and forest succession. Moreover, I have developed a quantitative model of the effects of light interception, and made qualitative predictions about heat and water, though I have not yet considered nutrients and biological interactions other than shading. Nevertheless, I have been surprised to find how much of nature's variation can be accommodated in a model that is still incomplete.

First it is necessary to define the problem. Chapter 2 develops a measure of the coverage of a canopy, analyzing the technique and its interpretation. Chapter 3 documents changes in canopy coverage during succession and measures the extent to which diverse species of trees prefer different degrees of canopy closure. Foresters have long been aware that the tolerance of the dominant species increases as succession proceeds, but part of their measure of tolerance is "that stage of succession at which the species is characteristically found." Hence I have developed a measure of tolerance that depends not on any a priori information about successional status, but only on a tree's relative ability to grow beneath a closed canopy. The confirmation of the foresters' axiom has raised further questions. For example, if tolerance is favored through succession, why are there any early stages at all? Why does succession not start with the most tolerant species?

To answer this and other questions, Chapter 4 creates a theory that distinguishes two extreme geometrical distributions of leaves in trees: a monolayer, with leaves densely packed in a single layer, and a multilayer, with leaves loosely scattered among several layers. In deep shade the lower leaves of a multilayer do not receive enough light for photosynthesis to balance their respiration. Conversely, in the open the multilayer can put out several layers of self-sustaining leaves to the monolayer's one. Thus the mono-layer grows faster in shade than the multilayer, but the multilayer holds an advantage in the open. Therefore multilayers should dominate the sunny canopy; but mono-layers, the shaded understory. The theory needs only a few comments about the distribution of heat loads among leaves to predict similar patterns of leaf distribution and morphology over climatic-edaphic gradients and at different stages of development in a single tree.

Chapter 5 recasts the theory in a more exact mold, forging mathematical expressions for the photosynthetic rates at different light intensities for trees with any number of layers and a variable density of leaves. These expressions justify the less rigorous predictions of Chapter 4; I also use them to create a model of succession. After discovering what kind of tree is best adapted to invade an open field, I simulate the environment in which the next generation of trees will grow and discover what morphology is best adapted to it. By repeating this process several times I can make realistic predictions about changes in the morphology of trees with succession, or I can construct hypothetical situations to answer questions of theoretical and practical importance. For example, is it possible to design a single distribution of leaves that can both reproduce continuously and compete effectively in its own shade? How does succession affect the total productivity of a forest?

Empirical tests of the theory require a measurement of just how many effective layers a given tree has. Such a measure is invented in Chapter 6, and it is used to confirm the predictions of the previous chapter in eastern deciduous forests of New Jersey, western coniferous forests in the Sierra of California, and mid-elevation rain forests in Costa Rica. The measurements also offer insights into patterns that are not explicitly predicted by the theory. For example, some early-successional trees are multilayered and so grow quickly in the open, but they shed such deep shade that they prevent their own reproduction as well as invasion by other species. Also, some late-successional monolayers are unable to reach the canopy for energetic reasons and so remain as a permanent understory in the climax forest.

Chapter 7 shows how the shape of a tree determines the total amount of light that it intercepts. The shape of a tree also constrains the distribution of leaves and is itself constrained by the tree's developmental pattern and successional status. These constraints are used to derive a table of optimal shapes and growth patterns for monolayered and multilayered trees.

Chapter 8 discusses issues that several helpful critics have raised, notably the origin and maintenance of adaptive patterns, adaptive compromises, and the uses of an incomplete theory. The chapter also summarizes the general implications of the theory, especially the relation of productivity, stability, and diversity to forest succession.

In order to make the arguments of each chapter as compelling as possible, I first make general but explicit assumptions and analyze them rigorously. Since some of the assumptions are patently unrealistic, I shall always discuss how relaxing them affects each result. Because I have approached the adaptive geometry of trees as a biologist rather than as a mathematician, I have presented my ideas and results in the order that seems most logical to my biological intuition. If the reader favors a more elegant and axiomatic approach and is looking for generality, he should start with Chapters 5 and 6, and plot his further course with the summaries at the end of each chapter.

CHAPTER TWO

Measurement of Light Intensity

The measurement of absolute light intensities in a sunlit forest is notoriously difficult due to the extreme range of intensities encountered and to their spatial and temporal variability. Furthermore, it is not clear how sunflecks and shaded spots should be averaged, from place to place or from time to time during the day, in order to give a figure that has physiological significance for plants. Many plant ecologists have avoided studying the role of light in forest succession because of the technical difficulties of making meaningful light measurements. Consequently the well-documented, analytical studies of vegetation are concerned with moisture and soil parameters that are easier to measure. Yet even in the reports of these studies the phrase "of course light also plays an important role" is almost universal.

Some technically minded ecologists have attacked the measuring problem with such elaborate equipment and such accurate sampling programs, that they have not had much time to study plants. The corresponding theoreticians have set the problem of comparing different plant communities, determining the amount of radiant energy that their photosynthetic structures intercept and the relation of this quantity to the productivity of the community. Thus their ambitious goal of an accurate, absolute measurement is indeed appropriate. However, their publications (reviewed by Anderson, 1964b, 1966), by emphasizing how far they are from their goal, have discouraged plant ecologists whose need for accuracy is much more modest.

For comparisons within the same woods, a very simple measure is adequate for my study. I measure the intensity of light at a given point in the forest relative to that above the forest, without worrying about the absolute changes from time to time. The temporal pattern of light at different points above the forest is roughly the same over topographically flat areas of the same latitude and climate. Thus the relative amount of radiant energy reaching a given point in the forest is inversely correlated with the density of foliage between that point and the sky. Therefore comparisons of foliage densities within a single geographic area are equivalent to comparisons of light intensities.

Since the literature on techniques of light measurement is voluminous and filled with cautions against my kind of measurement, I shall describe my measurement and defend my choice of it.

LEAF PROJECTION AND ALTERNATIVE
LIGHT MEASUREMENTS

A simple measurement of foliage density is insensitive to variations in the spectral composition of light, but such sensitivity may not be necessary for studies of photosynthesis in terrestrial plants. The spectral composition of light on a shaded forest floor differs markedly from that above the canopy, though sunflecks preserve the spectrum of sunlight (Evans 1939, 1956; Coombe 1957). However, as we shall see, the transmisson spectra of leaves do not differ markedly for species that are characteristic of the canopy, the understory, or the forest floor. Figure 2.1 shows the spectral characteristics of leaves as filters, plotted by a Cary recording differential spectrophotometer. There is no qualitative difference between the spectra of understory and canopy species in the visible range, especially in the major action spectrum of photosynthesis, from 4500

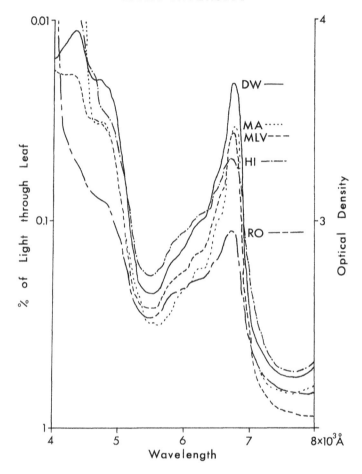

FIGURE 2.1. Transmission spectra of leaves from various heights in an oak-hickory forest.
Canopy: RO = Northern Red Oak, HI = Shagbark Hickory.
Understory: DW = Flowering Dogwood.
Shrub Layer: MLV = Maple-leaf Viburnum (*Viburnum acerifolium*).
Ground cover: MA = May Apple (*Podophyllum peltatum*).

The optical densities are two orders of magnitude higher than the absorption spectra published by Loomis (1965), for example, because the Cary instrument measures all light that is transmitted directly but only a small fraction of the light that is scattered or re-radiated within the leaf.

10

Å to 7000 Å (Lundegårdh 1966). The only difference is a tendency for the understory leaves to have a higher concentration of chlorophyll relative to other pigments than the canopy leaves. This can be interpreted as an adaptation to lowered light intensities in the absorption range of chlorophyll, around 6800 Å, rather than to changed light quality. One interesting difference between the spectra of canopy and understory leaves is the higher absorption of ultraviolet light by canopy leaves, and by Mayapples, which form a canopy near the ground in early spring in oak-hickory forests of New Jersey. Perhaps this is sunburn protection, but that is another problem. Although the light directly absorbed by chlorophyll is not the only light used in photosynthesis and the action spectra of other growth processes are quite different from that of photosynthesis, there is certainly no differential pattern of adaptive pigments among large terrestrial plants that equals the pattern of marine algae growing under different qualities of light at different depths (Haxo and Blinks 1950, Holmes 1957).

There is a second, and far more important, difficulty in interpreting foliage densities as physiologically significant light measurements. The amount of foliage overhead may be correlated with the amount of root tissue below the ground; the amount of foliage is certainly correlated with the water requirement per unit of ground area. This second difficulty remains even if light intensity and spectral composition are measured directly, because they are related to the amount of foliage above. Thus our interpretations are limited much less by the inaccuracies of light measurement itself than by the correlation of any light measurement with competition for both light and water by leaf and root.

I measure foliage density photometrically, either directly using a machine described below, or from photographs of the canopy. Using either method, I measure the amount

11

of sky obscured by foliage in a cone subtending 10° directly above saplings of various species. Of course the physiologically significant light does not come from directly above except near noon at low latitudes, but I assume that the measurement I am making is correlated with the amount of shading along the sun's average path during the growing season. This assumption is tenable only beneath a relatively uniform canopy, and I limit my study accordingly. I am also limited to studying virtually closed canopies, where more light comes from near the zenith than from low angles, rather than earlier stages of succession such as open fields.

The direct measurements are made with a cadmium sulphide photoresistor (RCA 4413) that has a narrow spectral response centered at 6000 Å. The rest of the physical machinery is shown in Figure 2.2. I set the sensitivity of the meter so that unobstructed sky at the zenith registers as full scale. Then I go into the closed parts of the woods and read a calibrated percent of unobstructed sky above saplings, checking the full scale reading from time to time against open sky lest a change in the ambient light levels affect the measurements. When the amount of light at the zenith is changing rapidly as it is near dawn, near noon, and in the late afternoon, as well as when the sky is unevenly cloudy with wisps of cirrus or scudding cumulus, the calibration cannot be made quickly or often enough, and the light meter is useless. This feature of the machine allows ample time for meals and, in the tropics, for a midday siesta. On very clear and sunny days, when the sky is deep blue and thin leaves are in direct sunlight, or if there is appreciable reflection of sunlight from shiny leaves or needles, the readings are unreliable. The readings are otherwise reproducible. Hazy and uniformly overcast days are comparable.

The alternative measurement involves photographing

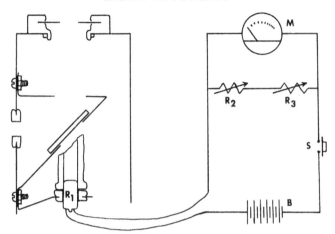

FIGURE 2.2. Light meter for measuring canopy coverage.

All parts are enclosed in a Bud "minibox." The layout of optical parts is shown on the left. Light from above is admitted through an iris diaphragm from an old camera, and passes through a baffle which supports a piece of a microscope slide which reflects part of the light through a ¼" i.d. rubber grommet to be viewed. Most of the light passes through to the photoresistor (R_1) which is mounted in another ¼" i.d. rubber grommet and shielded by an eyedropper bulb with the tip cut off. The innards of the case are painted flat black.

The rest of the electrical parts are shown on the right. B = 9V transistor radio battery. M = 100μA microammeter. R_1 = RCA 4413 CdS photoresistor. R_2 = 150Ω potentiometer for fine adjustment of full scale setting. R_3 = 1000Ω coarse adjustment of full scale setting. S = normally open push-button switch.

The scale is calibrated to % of light intensity at full scale after the instrument has been built. The calibrated scale looks approximately logarithmic.

the canopy on Kodak High Contrast Copy film. The percent of unobstructed sky is then read from the negatives, using a machine made from the butchered parts of a Photovolt 501 A densitometer, a Photovolt 52 C paper chromatograph scanner, and a Society for Visual Education AAA filmstrip projector. This method is useful under some conditions when the light meter cannot be used, as when the sky is brightly but unevenly lit. It gives a permanent and public record, but it is far more tedious, and the field processing of the film, which is necessary to assure results with a film of narrow exposure latitude, is a major disadvantage. A possible modification of the procedure is to take the pictures in the field on Plus X film, which has a wider latitude, and to reverse them onto High Contrast Copy for further analysis. However, the narrow latitude of High Contrast Copy is offset by its greater sensitivity in the blue than the green; thus it often gives better results than Plus X on sunny days with a deep blue sky. For my personal taste in field work, the light meter has an average advantage over photography.

Some of my colleagues have urged me to use a more direct measure of light intensity. They suggest setting out a grid of light integrators in the woods to measure directly the light available for photosynthesis. Such integrators should ideally imitate both the action spectrum of photosynthesis and the quantitative response of photosynthesis to different light intensities. However, although the photosynthetic process in most leaves reaches 90% of its maximum rate at about 20% of full sunlight, none of the simple chemical integrators reviewed by Perry, Sellers, and Blanchard (1969) have this property. Perry *et al.* suggest the photobleaching of chlorophyll as a measure of photosynthetically available light, but they do not show that the photobleaching of chlorophyll has the same action spectrum as that of photosynthesis. Hopkins (1962) has described an integrating meter that uses the actual growth of *Chlorella* as a measure of the light available for photosynthesis. In fact the geometry of the container for the *Chlorella* could be changed to vary the parameters of the quantitative response of photosynthesis to light intensity. Hopkins' method, having exactly the response of photosynthesis, is appropriate for short-term manipulative experiments. However, like any integrated light measure-

14

ment, it can only be used for comparisons among sites where the light has been integrated over exactly the same period of time. Furthermore, as with any measure of light intensity, it cannot distinguish between light and water competition among plants in the understory of a forest.

In contrast to any direct measurements of light intensity, the photographic assessment of foliage density offers a means of separating the effects of root competition for water and shoot competition for light. Photographs are taken with a "fish-eye" or Hill lens, after the method of Anderson (1964a) or Madgwick and Brumfield (1969), and the average path of the sun can be plotted on these pictures. At high latitudes the sun never crosses the zenith. The amount of foliage near the zenith can be taken as a measure of the effect of competition for water or for root space. The amount of foliage along the sun's path can be taken as a measure of shoot competition for light (Figure 2.3). The two effects can be disentangled in woods that are sufficiently patchy such that the two measures are not well correlated. I have experimented with this technique, using a camera mounted in a wooden frame that has a magnetic compass and a spirit level. The technique is practical only on hazy or uniformly overcast days, because any variation in skylight taxes the exposure latitude of the film for a high contrast image. I have yet to accumulate enough data to test the technique on a biologically significant problem.

In any case, the complete separation of root from shoot competition is an ecologically unreasonable goal that can only engender futile controversy. Any measurement of light on the forest floor, no matter how accurate and physiologically appropriate, confounds root and shoot effects since each bit of foliage overhead demands water and nutrients from below and sustains the activities of the roots that get them. In fact the initial effect of shade on some

15

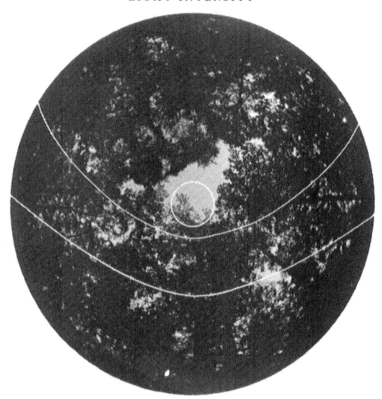

FIGURE 2.3. Fish-eye photograph of a gap in the canopy of an American Beech forest.
   The foliage in the central circle may be used as a rough estimate of water and nutrient demands at the spot where the picture was taken. The foliage in the curved band is in the sun's path during the spring and summer; thus it has a major effect on the light regime at the spot where the picture was taken.

plants is to decrease the relative growth of the root system and thus to lower the plant's tolerance of root competition (Eliasson 1968).

Therefore all measurements of foliage projection and percentage of unobscured sky that are used in the next chapter were made in a cone subtending about 10 solid degrees, by either photography or the light meter. I only used the light meter under conditions where I was confident that its readings were reliable, so that all measurements are in fact projections of sky or foliage in uniform stands and are unaffected by ambient light levels.

SUMMARY

Light is difficult to measure accurately because of spatial and temporal variations in intensity and quality. The variations in quality probably have little differential effect on photosynthesis because the leaves of most terrestrial plants absorb the same wavelengths of light. The relative light available for photosynthesis in two plants could be measured by simultaneously integrating the outputs of light meters whose curvilinear response imitates photosynthesis by leaves. However, the relative light available at some point on the forest floor depends on canopy coverage, because foliage and branches intercept incident light. Canopy coverage can be measured more easily than light itself, using a special light meter or photographic techniques. This measure is appropriate to compare the light environment at different points in the same woods. If the woods are uniform enough, measuring the canopy near the zenith alone is sufficient. If necessary, canopy coverage at any angle can be measured from "fish-eye" photographs. In particular the canopy coverage along the sun's path can be taken as a measure of competition for light and the canopy coverage near the zenith as a measure of water and nutrient demands on the ground below. However, root competition and shoot competition are not physiologically independent, much less ecologically distinct. I shall therefore limit my measurements to the canopy coverage directly above in a cone subtending 10 solid degrees. All measure-

17

ments are made in woods that are uniform enough that the coverage near the zenith fairly represents the coverage at lower angles. Since root and shoot competition are not distinguished, the total effect of the canopy is measured, rather than the effect of shading alone.

CHAPTER THREE

Analysis of a Forest Succession

Since light is critical to the growth of green plants, the shading of trees in the forest canopy should profoundly affect the growth of seedlings and saplings in the understory. If we know the pattern of shade beneath the canopy and the light required by saplings in the understory, we can predict which species of sapling should grow vigorously enough to become the next generation of the canopy. If none of the saplings have unusual requirements for minerals, moisture, and such, then our predictions should correspond to what we know about the history of a particular forest succession.

Foresters long ago stated the first axiom of the effects of shading on forest succession: species that are progressively more shade tolerant become dominant as succession proceeds toward climax. Unfortunately, the measurement of tolerance that foresters often use includes information about the stage of succession at which the species in question is characteristically most abundant. Thus when the axiom is examined critically, it is found to be either circular or unsupported, even though it is intuitively reasonable. A theory based on this axiom would allow prediction and perhaps management of forest succession, as well as the design and construction of particular shade environments. A theory of the effects of shading on forest succession would also provide a standard against which to measure the effects of other factors.

Therefore the statement of the problem is quite simple, and the needed data are also simple: a measure of the amount of shade that each species casts when it is in the canopy, and a uniform measure of the shade tolerance of saplings of different species. Ideally the problem is one of sampling points in a forest randomly, measuring the light intensity under each kind of tree and the light intensity at which each species of sapling is found.

Despite technical and practical difficulties, I have measured the shade cast by different species in the canopy, and the ability of different species of saplings in the understory to grow beneath canopies that shed different amounts of shade. These measurements confirm the axiom that tolerance increases with succession, but only if tolerance is defined to include effects of root competition as well as shade. The measurements also allow the discovery of other factors, besides root and shoot competition, that affect succession. Finally the measurements suggest new questions and new ways to answer old questions.

STUDY AREA AND METHOD

The area that I have studied most intensively is the woods behind the Institute for Advanced Study in Princeton, New Jersey. This woods was farmed in blocks that have been left fallow for varying amounts of time. Throughout its history there have always been many successional stages present locally; thus most of the tree species that are characteristic of different stages of succession have had access to each plot within the woods at all times.

Figure 3.1 is an aerial photograph of the woods with discrete stands mapped and the dominant species noted. The oak-hickory and Gray Birch forests are on well drained soil. The mean water table is well below 160 cm, as shown by the complete oxidation of the soil to that depth. The American Beech and Red Maple forests are poorly drained, with patches of reduced soil at a depth of

FIGURE 3.1. Aerial photograph of woods behind the Institute for Advanced Study in Princeton, New Jersey.

The photograph is reproduced with the kind permission of the Agricultural Stabilization and Conservation Service, United States Department of Agriculture.

90 cm. The Sweetgum and Blackgum are on soils that are moist in the spring but dry by the end of summer, reduced at a depth of 130 cm. The pine forest was not studied since it is a plantation. Neither was the flood plain forest studied because its canopy is very patchy, and many parts of the oak-hickory forest were also patchy enough to be ignored.

Successions for Piedmont New Jersey have been described by Bard (1952), Niering (1953), and Buell, Langford, Davidson, and Ohmann (1966). In view of these studies, the succession in the Institute Woods can be reconstructed as follows: a field is colonized by Eastern Redcedar or Gray Birch; later Red Maple, Sweetgum, or Blackgum invade; and finally these are replaced by American Beech on moist soils or by oak-hickory on dry. Buell *et al.* (1966) present evidence that Beech eventually replaces oak-hickory even in dry sites in northern New Jersey. The Beech and oak-hickory are treated in this study as the climax forest appropriate to moist and dry soils respectively. Whether they represent a true climatic-edaphic climax or not is irrelevant to my study and discussion. Both associations will clearly persist for a far longer time than any of the preceding stages, and there is no evidence that Beech would soon replace oak-hickory or vice versa in the Institute Woods without a change in the edaphic conditions.

Figures 3.2 through 3.5 show these stages of forest succession. The Gray Birch forest (Figure 3.2) is well beyond the field stage. There are many shrubs and saplings in the understory. Light penetrates the canopy copiously and uniformly. The Red Maple forest (Figure 3.3) is atypical in one respect; the dominant trees are stump sprouts from a previous cutting, rather than direct invaders. Thus the canopy is slightly more uniform than in a stand of uneven aged trees. The understory is typical, however, with nu-

FIGURE 3.2. Gray Birch forest, Institute Woods, Princeton, N. J.

FIGURE 3.3. Red Maple forest, Institute Woods, Princeton, N. J.

FIGURE 3.4. American Beech forest, Institute Woods, Princeton, N. J.

FIGURE 3.5. Oak-hickory forest with understory of Flowering Dogwood, Institute Woods, Princeton, N. J.

merous seedlings and saplings of many species. Less light reaches the understory than in the Gray Birch forest, but again its pattern is uniform, with many tiny sunflecks. The American Beech forest (Figure 3.4) contrasts in all respects; the only saplings in the understory are a few Sugar Maples and the root sprouts of Beech. The scant light reaching the forest floor comes in large patches through distinct gaps in the canopy. The oak-hickory forest has an understory of Flowering Dogwood (Figure 3.5), with virtually no saplings; the only common plants below the Dogwood are *Viburnum,* Mayapple (*Podophyllum peltatum*), and assorted ground cover. The light on the forest floor is scant and patchy as in the Beech forest, but a large part of the shade is cast by the Dogwoods. The paucity of saplings in a young stand suggests that the oak-hickory forest had been used as woodland pasture, but the shade is now deep enough to retard further invasion. Figure 3.6 summarizes the dramatic changes in the canopy during succession.

My method was to measure the proportion of unobscured sky directly above each sapling in the woods, and to note the species of tree in the canopy directly above it. From these notes for a uniform woods I constructed histograms of the amount of shade cast by each species of canopy tree, and the extent of the canopy above vigorous saplings of each species.

Because of my method of sampling, the histogram of the shade under various species in the canopy is biased; I do not have a random sample of a given woods or a representative sample of the light levels to be found under each species. Rather, I have a sample of light intensities at which saplings of any species grow under each species in the canopy. Thus my sample is not random, but it is appropriately biased to represent the relative amounts of canopy closure under which saplings are found.

A. Gray Birch forest.

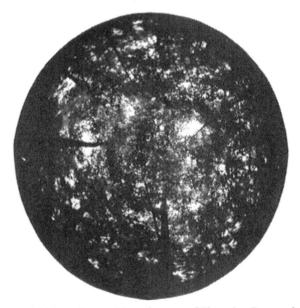

C. Oak-hickory forest with understory of Flowering Dogwood.

FIGURE 3.6. Fish-eye photographs of canopies in the Institute Woods, Princeton, N. J.

B. Red Maple forest.

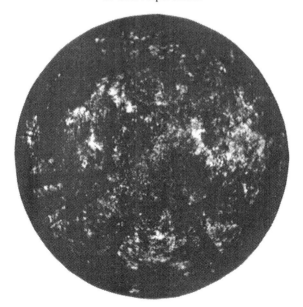

D. American Beech forest.

FIGURE 3.6.

The distribution of shade levels at which each species of sapling grows is subject to a much more serious bias. I have not sampled all shade conditions equally; therefore the raw histograms cannot provide information about the absolute canopy preference of the species in question. Within a single, uniform stand of trees, I can of course make valid comparisons of the preferences of different species, but as soon as I move to a different area, I cannot make cross-comparisons, because the range of possible shade conditions differs from one stand to another.

RESULTS AND INTERPRETATIONS

A preliminary analysis of the amount of sky unobscured by various canopies showed very different mean values for canopies composed of different species, but the coefficients of variation appeared to be uniform. The distributions are skewed since of course no measurement could be less than zero. Therefore a logarithmic transformation was used to make the variance more uniform and the distribution more symmetrical (Sokal and Rohlf 1969). There are theoretical as well as statistical reasons for this transformation. As I shall show in Chapter 4, the proportion of open sky is an exponential function of the actual density of leaves if the leaves are distributed independently with uniform density. Therefore the local density of leaves is more closely related to the logarithm of the proportion of open sky than to the proportion itself. Empirically, the transformed data have a similar variance for each species. Thus I need only give the mean to specify the distribution of proportions of light through a canopy of each species.

The mean percentage of incident light that penetrates various canopies, and parameters of its distribution, are listed in Table 3.1. The ratio of mean to variance for the transformed data is a measure of the evenness of the

TABLE 3.1. Percentage of skylight that penetrates canopies of various species. The mean % of skylight through the canopy is followed by the larger of the deviations found by adding or subtracting the standard error of the mean of the log-transformed data. The evenness of the canopy is measured by the ratio of mean to variance of the logarithm of the proportion of skylight penetrating the canopy.

Canopy	% of skylight through canopy		Evenness of canopy	Number of meas- urements
Gray Birch	20	±5.0	2.6	13
Bigtooth Aspen	19	±4.0	2.1	18
Sassafras	15	±2.5	2.4	27
Tuliptree	11	±1.7	2.9	32
red oak	10	±1.0	3.2	68
Sweetgum	8.6	±1.2	2.8	48
Red Maple	7.8	±0.5	2.2	293
White Oak	7.7	±1.4	3.3	25
hickory	7.0	±1.5	4.2	13
Black Oak	7.2	±0.7	7.3	38
Blackgum	6.9	±0.6	4.9	65
American Beech	3.1	±0.4	4.8	49
oak, hickory; Tuliptree with Flowering Dog- wood understory	2.9	±0.3	7.3	32

canopy; the more evenly leaves are distributed within the canopy, the lower the variation from sample to sample, and the higher the ratio of mean to variance. From Table 3.1 it is obvious that the shade cast by the canopy and the evenness of the canopy both increase as succession proceeds. The horizontal density of leaves within the canopy increases and becomes more homogeneous, even though the canopy itself may become more patchy (cf. Figures 3.2–3.6).

What effect does shading have on saplings? Because the canopy measurement confounds the effects of shoot and root competition, I shall refer to a canopy effect, as explained in Chapter 2, rather than a light effect.

A simple and uniform measure of growth is a year's elongation of the terminal twig of a sapling. This growth can be measured as the distance between bud scars of successive years, and when yearly growth is plotted against the amount of canopy overhead, the curves of Figure 3.7 result. The variance is large, but trends are clear: it is immediately obvious that the canopy has a deleterious effect on growth; that the canopy measure that I have made is a significant one, albeit with large variance, for the growth of saplings. Furthermore, the deleterious consequences of shading become apparent at shade levels that I have found under several species of trees in the canopy.

I now return to the problem of determining canopy preferences for different species of saplings. I should be able to use histograms of those canopy projections under which each species of sapling is found; however, to do this blindly would produce a serious sampling bias. Suppose, for example, that data are pooled from a large and heterogeneous area. American Beech sprouts are most numerous under Beech trees and Red Maple saplings under Maple trees. Since I have already shown that the shade under Beeches is deeper than that under Maples, I would of course find the Beech sprouts under more shady conditions than the Maple saplings.

An equal sampling of all possible canopy projections is impossible, but I can at least sample all species of saplings from the same distribution of canopy projections. As long as comparisons are made within a single, uniform stand, there are no problems. If I pool data from different stands, recalling that the shade under each species of tree has a characteristic distribution, then I can construct the histogram of canopy projections at which saplings of different species are found under one kind of canopy. For example, I look at the relative canopy preferences of different species of saplings when they are allowed to "choose" from the distribution of canopy projections under Red Maple trees.

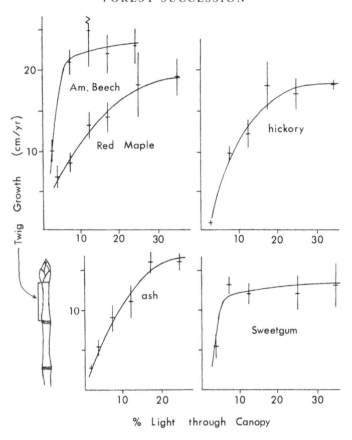

FIGURE 3.7. Effect of canopy on annual growth of twigs.
Means are plotted with a bar extending to ± the standard error
of the mean. Smooth curves are fitted by eye.

The canopy preferences of several species under Red
Maple are presented in Table 3.2. Gray Birch and Bigtooth Aspen are never found under a Red Maple canopy,
though they are often found in close association with Red
Maples. Sassafras, Sweetgum, and ash are found when at
least 80% of the sky is obscured by a Red Maple canopy.
Blackgum, hickory, and Red Maple itself are found under

TABLE 3.2. Canopy preferences of various saplings under Red Maple. The canopy preference is the log-transformed mean percent of skylight penetrating the Red Maple canopy above a given species of sapling. It is followed by the larger deviation found by adding or subtracting the standard error of the mean of the log-transformed data. Tolerance is taken from Baker's (1950) table.

Species	Canopy preference (% skylight)		Tolerance	Number of measurements
Gray Birch	100		very intolerant	15
Bigtooth Aspen	100		very intolerant	7
Sassafras	15	±2	intolerant	26
Blackgum	14	±2	tolerant	18
hickory	12	±2	intolerant	61
Sweetgum	11	±2	intolerant	32
ash	10	±2	intermediate	36
Red Maple	9.5	±1.4	tolerant	39
American Beech	7.0	±1.0	very tolerant	46
Black Oak	6.8	±0.8	intermediate	13
Flowering Dogwood	5.7	±1.5	very tolerant	40

slightly more dense canopies, and Flowering Dogwood, American Beech, and oak are found under canopies that cover more than 90% of the sky.

If the "tolerance" of foresters is defined in its most general sense as the ability to grow beneath an overstory (Decker 1952), then my "canopy preferences" should be objective and analytical measures of tolerance. In fact the canopy preferences of Table 3.2 correlate quite well with the tolerances intuitively assigned by foresters. The only exceptions are Blackgum and Black Oak. Indeed I have found Blackgum growing under much shadier conditions in Mammoth Cave National Park, Kentucky. For Black Oak I have no explanation.

Furthermore the measurement of canopy preference is not itself dependent on successional stage, and it therefore provides an independent test of the axiom that the tolerance of the dominant species increases as succession approaches climax. The most intolerant species, Gray Birch and Bigtooth Aspen, are indeed pioneers in succession, while the most tolerant species, oak, American Beech, and Flowering Dogwood, are components of the climax forest.

Also note that the most tolerant tree is the one that ultimately produces the densest canopy, Beech, or oak with a Dogwood understory. If tolerance as a sapling and shade cast by the canopy were not correlated in late-successional species, then the most tolerant species would regenerate best to form an open canopy, which would allow growth by less tolerant forms, which would in turn slow the regeneration of the more tolerant species. The outcome would be difficult to predict — perhaps a locally cyclic mixture of the two species at climax.

There is no reason to believe that the tolerance orders that I have measured at a particular stage of succession, namely the stage dominated by Red Maple, are the same at any other stage or under other edaphic conditions. In general, however, the canopy preferences under Red Maple should correspond to tolerances in the edaphic conditions that are locally appropriate to a Red Maple stand, namely moist soils, while the canopy preferences under oaks should correspond to tolerances in the well-drained soils appropriate to an oak stand. I therefore plot the canopy preferences of different species of saplings under various canopy species in Figure 3.8. Several interesting patterns emerge.

Under a variety of canopies American Beech is more tolerant than Red Maple, which is in turn more tolerant than Blackgum. This suggests a common successional series in which Blackgum is replaced by Red Maple, which is replaced by Beech. I shall return to this pattern in a moment.

Under an oak canopy, hickory is as tolerant as Beech. Recall that hickory is a component of the climax forest under the edaphic conditions appropriate to an oak over-

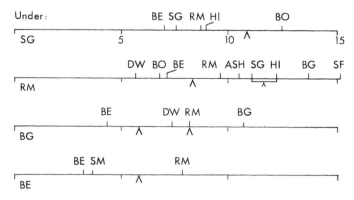

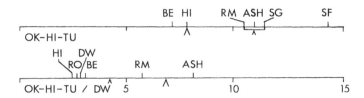

FIGURE 3.8. Canopy preferences of saplings under a variety of canopies.

For each species I have plotted the log-transformed mean % of skylight reaching saplings through the given canopy. The arrows separate groups of means that are different at the 0.05 level by a 2-tailed t-test (Sokal and Rohlf 1969). The soils beneath the oak-hickory-Tuliptree canopies are drier than those beneath the other four canopies.

ASH = ash.   BE = American Beech.   BG = Blackgum.
BO = Black Oak.   DW = Flowering Dogwood.   HI = hick-
ory.   OK = oak.   RM = Red Maple.   RO = red oak.
SG = Sweetgum.   SF = Sassafras.   SM = Sugar Maple.

story. Thus the axiom that tolerance increases as succession proceeds is confirmed (or at least not denied), even for changed edaphic conditions, as long as the measurement of tolerance is made under the appropriate edaphic conditions.

Finally, Sugar Maple seedlings are found under a Beech

canopy as dense as that where Beech root sprouts are found. Root sprouts have a supply of energy other than their own leaves so that Sugar Maple may actually be more tolerant than Beech even though it sheds less shade. Ward (1956) has adopted this tolerance order to explain the slow geographic replacement of Beech by Sugar Maple in Wisconsin. If this tolerance order is borne out by further data, I should have an example of two species late in succession whose tolerance and shade shed are inverse. Hence it is interesting to note that American Beech and Sugar Maple are co-dominant in a mixed climax association over a large part of the northeastern United States.

DISCUSSION

Most of the problems that I have set can be examined with great quantities of the kinds of data I have gathered. A diagram of the percent of sky unobscured by the canopy versus the canopy preference of saplings in the understory allows prediction of the course of succession from those tolerance orders that are consistent. Inconsistencies point to edaphic conditions that are inappropriate for a particular successional series. The diagram summarizes much of the data that are needed to predict and manage succession on a given site.

The particular prediction that Red Maple should succeed Blackgum and be replaced by American Beech, gives a more detailed order of replacement than that found in previous studies of succession in central New Jersey. However, the Institute Woods has stands of Blackgum, Red Maple, and Beech, each stand containing all of the other species. It is possible therefore to determine whether Red Maple and Beech are indeed invading the Blackgum stand, whether Beech is invading the Red Maple stand, and whether the Beech stand is being invaded at all. I measured the diameters at breast height of all trees in a sample of each stand. Within each species the diameters are correlated with age, and the distribution of diameters within a stand measures the relative successional status of a species (Daubenmire 1968). A large number of seedlings but no saplings, poles, or large trees denotes an unsuccessful invasion. A large number of seedlings, several saplings, and a few poles constitute an invasion. Many seedlings and progressively fewer saplings, poles, and large trees indicate a population that is continually replacing itself. Very few small trees and many large ones represent a population that is not continually replacing itself; its successional status depends on the status of other trees in the stand. If other species are invading or continually replacing themselves, then a species with a senile age distribution has seen its day. However, if there are no other species regenerating more effectively, then very little regeneration may be sufficient to assure the persistence of a species with a senile age structure.

The size distributions of trees in the Blackgum, Red Maple, and American Beech stands of the Institute Woods are presented in Table 3.3. As predicted from their relative tolerances, Red Maple is invading the Blackgum stand, Beech is invading the Red Maple stand where Blackgum is senile, and there is no significant invader in the Beech stand.

The measurement of tolerance can also be used to interpret other patterns of competition for space between plants. Departures from the predicted patterns may then suggest which factors other than root space and shoot space must be considered to interpret the distribution of a species. For example, I have already demonstrated a range of tolerance among several species. Thus I predict that as the canopy becomes denser, the less tolerant species should be eliminated and the diversity of saplings should decrease. Of course under very open canopies the shade-requiring

TABLE 3.3. Relative age structures of three stands in the Institute Woods, Princeton, N. J. The basal area of each species is given as parts per million of ground area. The area sampled and age of oldest tree are: Blackgum woods = 550 m², 60 yrs; Red Maple woods = 759 m², 98 yrs; American Beech woods = 2236 m², 210 yrs.

Species	Basal area (ppm)	Midpoint of each diameter class (cm)						
		0	5	15	25	35	45	>50
BLACKGUM WOODS								
Blackgum	828	43	38	5	4	1		
Bigtooth Aspen	484		2	11	1	1		
oak	224	37	9	1		1		
Red Maple	113	166	37	2				
Gray Birch	104		3	2				
Tuliptree	72	14			1			
American Beech	2	19	2					
other species	86	128	18	2				
RED MAPLE WOODS								
Red Maple	2154	43	14	6	6	3	4	1
oak	683	6		3	3	2		
Sweetgum	146	8	1			1		
American Beech	53	18	8	2				
Blackgum	11	17	17					
other species	30	167	46					
AMERICAN BEECH WOODS								
American Beech	2823	57	217	3	3	3	10	14
Tuliptree	516		17	13				2
Blackgum	214	4	3					2
White Oak	97	1	1					1
Red Maple	45	8	13			1		
other species	37	38	61	1				

species may be eliminated. The number of tree species in the understory, therefore, should be largest beneath canopies that shed some intermediate amount of shade and should fall as the canopy becomes denser or lighter. In Figure 3.9 I have plotted the diversity of saplings under canopies of different densities. The diversity of saplings is measured as $1/\Sigma p_i^2$ where p_i is the proportion of saplings of species i in the sample of all saplings found under each canopy. This is Simpson's (1949) measure of diversity; it counts the species, weighing each by its abundance. Figure 3.9 agrees with my prediction, except that the diversity of saplings under Sassafras is much lower than would be expected on the basis of the density of the Sassafras canopy. My prediction is confirmed, and the conspicuous exception stands out in ways other than its departure from

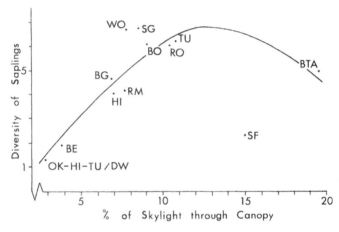

FIGURE 3.9. Diversity of saplings under canopies of various densities.
    Diversity is measured as $1/\Sigma p_i^2$, where p_i is the proportion of saplings of species i in the sample of all saplings found under each canopy. It is the number of equally abundant species for which the probability of withdrawing two conspecific individuals at random is the same as it would be for the actual distribution of species. The line is a qualitative prediction based on the effects of shading alone.
    BE = American Beech.   BG = Blackgum.   BO = Black Oak.   BTA = Bigtooth Aspen.   HI = hickory.   OK-HI-TU/DW = oak, hickory, Tuliptree, and Flowering Dogwood.   RM = Red Maple.   RO = red oak.   SG = Sweetgum.   SF = Sassafras.   TU = Tuliptree.   WO = White Oak.

the prediction. It is the only tree in Figure 3.9 with highly aromatic leaves. When I first plotted the figure, I guessed that Sassafras might exude some chemical that is detrimental to other species of trees. To test this guess, I collected several species of dry leaves that had lain on the forest floor for 8 months. I ground them in a Waring Blender, left 2 gm of chopped leaves in 50 ml of distilled water overnight, and filtered the resulting tea through Whatman #1 filter paper. Then 5 ml of the tea was poured on filter paper in a 10 cm Petri dish for 48 hr germination tests of Northrup-King Grand Rapids Lettuce seeds. The results of these tests are shown in Table 3.4. Extract of

TABLE 3.4. 48-hour germination tests with Lettuce seeds in 10-hour water extracts of various dry leaves.

Extract	Germinated	Not germinated
Sassafras	4	46
Sweetgum	35	15
White Oak	24	26
Southern Red Oak	28	22
Red Maple	29	21
Distilled water	35	15

Sassafras clearly inhibits germination of Lettuce seeds to a far greater extent than the extract of any other species. This fact proves only why one seldom finds Lettuce under Sassafras trees, but it also suggests further experiments to discover which species of trees are inhibited by the chemistry of Sassafras. It is worth noting that walnut trees have a well known and potently herbicidal exudate. The pattern of a tree with sparse foliage and a dearth of saplings beneath is very striking for Black Walnut and Butternut trees in closed woods (Grummer 1961). Thus the measurement of the shade cast by the canopy and the diversity of saplings below provide a rapid survey for interactions that are unrelated to light.

Several venerable questions are left unanswered by simple measurements of the shade cast by the canopy, but these questions can now be phrased in terms of such measurements and asked in a more rigorous fashion. Why does the amount of shade change gradually through succession, rather than increasing suddenly to its final value soon after trees first invade a field? Do trees that are tolerant of a closed canopy necessarily shed deep shade themselves? If tolerance is favored through succession, why are there any early stages in succession; why does succession not start with the most tolerant species? Is it possible for a species to reproduce continuously and compete effectively in its own shade? These are the questions for which I shall propose a theory. To the extent that I have tested the theory, it partially answers them. The untested parts of the theory suggest the kinds of data that will eventually answer these questions completely.

SUMMARY

Since the trees that invade late in succession must grow up from the shaded understory, we predict that species that are progressively more tolerant become dominant as succession proceeds. To confirm this prediction we need a measure of the amount of shade cast by the canopy at each stage of succession and a measure of tolerance that is independent of successional stage. The former measurement is simply the average (geometric mean) proportion of skylight that penetrates the canopy, while the latter measurement is ideally a similar average of the proportion of skylight that penetrates the canopy above each species of sapling in the understory. In this study, I made such measurements in woods of different successional stages in central New Jersey.

Geometric means are used since the amount of light that penetrates foliage is closer to a negative exponential function of leaf area than to a linear function. The shade cast by the canopy and the evenness of the canopy both increase as succession proceeds. Under dense canopies the annual linear growth of twigs of several species falls markedly as the canopy becomes more dense.

The relative canopy preference of a species under a given canopy is defined as the geometric mean of the light that penetrates the canopy above a sample of saplings of that species. Canopy preferences correlate well with the tolerances traditionally assigned by foresters. Canopy preference is then a measure of tolerance that includes no a priori information about successional status. The successional status of a species can be guessed from its age structure relative to other species in its stand, and measured accurately if stands of different ages are compared. With these independent measurements of successional status and tolerance, the axiom is confirmed: the tolerance of the dominant species increases as succession proceeds. The axiom even holds for different physical environments, as long as the measurements of tolerance are made under appropriate edaphic conditions. The most tolerant trees are apparently those that shed the deepest shade. If tolerance and shade cast are not correlated for trees that are late in succession, several species may persist in the climax forest.

The number of species in the understory should decrease as the canopy becomes more dense, unless the canopy is open enough that tolerant species are parched. The actual diversity of saplings behaves accordingly, being high under canopies that shed an intermediate shade and low under very dense or very sparse canopies. A conspicuous exception is the low diversity of species under Sassafras, which sheds an intermediate amount of shade. There is something in the leaves of Sassafras that inhibits germination of Lettuce seeds, and this factor may be implicated in the low diversity of seedlings under Sassafras trees.

Tolerance has been measured as an effect of a canopy rather than as an effect due to light alone. All of the measurements have been made in a single locality. Despite the lack of physiological precision and global generality, the results allow several general questions to be set precisely. Do tolerant trees necessarily shed deep shade? Can a single species be a tolerant pioneer? Is it possible for a species to reproduce continuously and to compete effectively in its own shade?

CHAPTER FOUR

Theoretical Strategies of Leaf Distribution

I shall start the theoretical inquiry by presenting three seemingly unrelated bits of information, and then use them to generate a table that suggests optimal parameters and adaptive advantages of two different spatial distributions of leaves in plant communities. The distributions are: a monolayer, with leaves densely packed in a single layer; and a multilayer, with leaves loosely scattered among many layers. A monolayer can obviously expose no more than one unit of self-sustaining leaf area for each unit of ground area. A multilayer, however, can expose much more leaf area, and its leaves pay for themselves if they are small and far enough apart such that no leaf completely eclipses the sun from any leaf in the next layer. To find out just how much more leaf area a multilayer can put out, how small the leaves must be, and how shade affects the strategies, we must first determine the pattern of shade behind a single leaf, the photosynthetic response of a leaf to different light intensities, and the amount of light that penetrates a random distribution of leaves.

THE SHADOW OF A LEAF

Many theoretical studies of the light climate of woodlands assume that the sun is a point source that moves in a semicircle through the zenith. I shall initially make the equally absurd assumption that the sun is a source of finite size that doesn't move. A ball of light 8.6×10^5 miles in diameter and 9.3×10^7 miles away is completely blocked by an opaque circular leaf of diameter d for a distance of about $108d$ (Figure 4.1). Suppose now that we have two layers of leaves. If the second layer is closer than 108 leaf

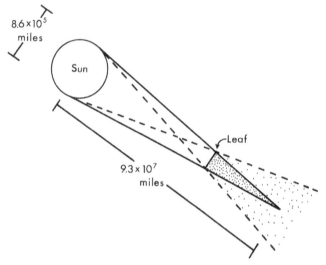

FIGURE 4.1. Shadow of a sunlit leaf.
   The scale is obviously distorted, but the triangle including the sun is similar to the triangle containing the leaf. Hence (length of umbra)/(distance from tip of umbra to sun) = (diameter of leaf)/ (diameter of sun). Length of umbra = (9.3×10^7)(diameter of leaf)/(8.6×10^5) = 108 (diameter of leaf).

diameters to the first, parts of some leaves in the second layer are completely shaded (Figure 4.2, layer A). If the second layer is farther than 108 leaf diameters from the first, then all parts of all leaves in the second layer are lit by at least part of the sun (Figure 4.2, layer B). More generally, a leaf layer blocks the sun completely in some places up to 108 leaf diameters away, but from much farther away the effect of a leaf layer is essentially that of a neutral density filter.

46

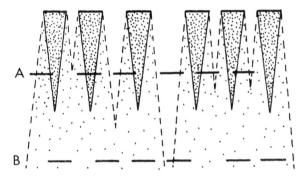

FIGURE 4.2. Shadow of a leaf layer.
Leaves in the first layer would completely obscure the sun
from parts of leaves in a layer at A, but not from any part of
a layer at B.

A few comments are in order about the reality of my assumptions and the effect of their falsity on the result. A
ball of light 8.6×10^5 miles in diameter and 9.3×10^7
miles away would subtend an arc of about 0.5 solid degrees.
The brightly lit aureolus around the sun subtends much
more; thus the figure of $108d$ should be less. Empirically
the distance at which the shadow of a circle vanishes is
about 50 to 70 times its diameter with the sun at the zenith
on a clear day. On a uniformly cloudy day, light comes
from 180 solid degrees, and the shadow of a circle vanishes
at a distance equal to its diameter. Thus the average length
of the umbra of a leaf is much greater in sunny than in
cloudy climates, but it is still proportional to the diameter
of the leaf. Therefore if no leaves are to be subjected to
total shade, the minimum distance between layers of a
multilayer is very large in sunny climates, and significantly
smaller in cloudy climates. Correspondingly the size of a
leaf for a given distance between layers is very small in a
sunny climate, larger in a cloudy climate.

Leaves are generally not circular in shape; since we are
concerned with the largest circular area of the leaf that

completely blocks light, the effective diameter is simply the diameter of the largest circle that can be fully inscribed within the leaf outline. For example, a thin, elongate leaf of pine or a deeply lobed oak leaf has a smaller effective diameter than a round or elliptical leaf of equal area.

Leaves are not opaque. However, the characteristics of the average leaf as a filter include an extinction factor greater than 95% per leaf in the absorption range of the chlorophylls and carotenoids (Figure 2.1 in Chapter 2, and Loomis 1965).

In summary, beyond a certain distance, which depends on the diameter of the convex portion of the leaf, a layer of leaves acts essentially as a simple density filter, reducing the light intensity by the proportion of flux intercepted by the leaves.

PHOTOSYNTHETIC RESPONSE TO LIGHT

What is the efficiency of photosynthesis at different light intensities? There is a great deal of variation from species to species. Much of this variation will eventually be explained as due to the geometry of leaf placement, which results in a variable amount of mutual shading. However, when the net amount of photosynthesis per unit of lit leaf area is plotted for different light intensities, a general pattern emerges (Figure 4.3). A certain amount of light, about 2–3% of full sunlight, is necessary for photosynthesis to balance respiration. Above this compensation intensity, net photosynthesis increases with increasing light intensity until the light-trapping and energy transfer mechanisms of the plant are saturated; that is, until the rate at which light is trapped is no longer the limiting factor in photosynthesis. This saturation point must represent a fundamental limitation of the photosynthetic mechanism of terrestrial plants, because the light intensity at which the mechanism becomes saturated, other things being equal,

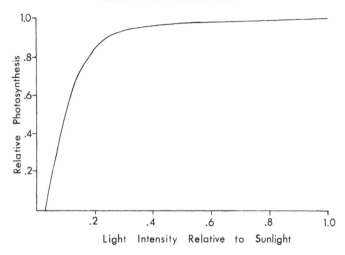

FIGURE 4.3. Typical effect of light on the net photosynthesis of a fully illuminated leaf (after Baker 1950).

is in the same range for a wide variety of plants, namely about 20% of full sunlight (Büsgen and Münch 1929, Baker 1950).

The first and most important implication of this saturation is that it defines the conditions when strategies of light interception are important. As long as there is a sufficient supply of moisture and nutrients to support enough vegetation to cut the average intensity of light at ground level to less than 20% of full sunlight, then light is among the limiting factors for seedlings in that community. Although light need not be the only, or even the most important, factor limiting the growth of plants, the relative strategies of light interception can always be superimposed on other strategies when the amount of leaf area relative to ground area is more than 80%.

A simple example will outline the theory to come. Since the curve of net photosynthesis saturates at 20% of full sunlight, any leaves that are lit more intensely must operate at

49

full capacity. Thus we can imagine a layer of leaves with one unit of leaf area for every 2 units of ground area, operating at the peak rate in full sunlight (Figure 4.4). About 70 leaf diameters below this layer, the light intensity is reduced to 50% of full sunlight, and we can imagine

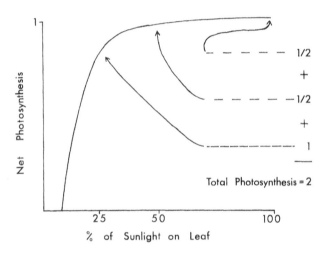

FIGURE 4.4. Photosynthesis of several layers of leaves. See text for further explanation.

a second layer of leaves, one leaf for every 2 units of ground area, again operating at peak photosynthetic rate. Finally, about 70 leaf diameters below this second layer, the light intensity is 25% of sunlight, and since this is just above the saturation point for photosynthesis, we can cover the available ground area with a third layer of leaves, all operating at peak photosynthetic rate. Thus we design a multilayered strategy that exposes ($\frac{1}{2}$ + $\frac{1}{2}$ + 1 = 2) twice as much leaf area as there is ground area, with all leaves operating at peak photosynthetic rate. On the other hand, a monolayered strategy can never do better than to cover completely the available ground area with a single layer of leaves.

Of course, no tree species is developmentally capable of our ideal multilayered strategy (although some communities as a whole might approach it) because it requires subtle and exact gradients in leaf spacing over a considerable vertical range. Thus we have to examine the actual distributions of leaves that might be generated by constant leaf density.

PROJECTION OF RANDOM LEAVES

Again I take circular leaves of radius r, and I put them out randomly in a single layer at an average density of ρ leaves for each unit of ground area (Figure 4.5 A). If I now take another random point and ask whether or not it is covered by one or more leaves, this is the same as asking whether or not there is a leaf whose center is within a distance r of the new point (Figure 4.5 B). Since the leaves are randomly distributed, the Poisson probability that there are no leaves within a distance r of a random point, i.e., within a circle about that point of radius r and area πr^2, is $\epsilon^{-\rho\pi r^2}$. (See any elementary statistics text or MacArthur and Connell 1966. The same distribution is derived from different assumptions by MacArthur and Horn 1969.) If we take a large number of sample points, then, the average proportion of points not covered by leaves is also $\epsilon^{-\rho\pi r^2}$. The projection of leaves is defined as the proportion of ground covered by the shadow of the leaves when they are illuminated by an infinitely distant point of light; hence it is the proportion of light flux that they intercept. Therefore the projection of the leaves is the average proportion of sample points covered by one or more leaves, which is equal to

$$1 - \epsilon^{-\rho\pi r^2}. \qquad (4.1)$$

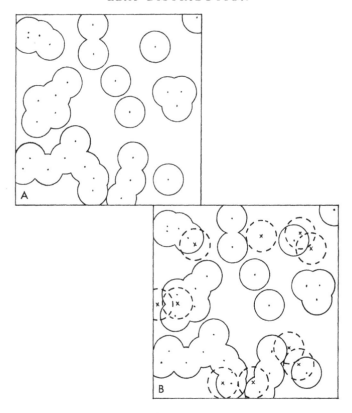

FIGURE 4.5. Distribution of leaves.
A. 30 circular leaves with their centers randomly placed on a
20 x 20 square grid.
B. The same as A, but with 10 randomly placed samples (dotted
lines), each a circle about the point X with the same radius as
the leaves.
  The sample around each X that is covered by a leaf contains the
center of one or more leaves; that around every uncovered X,
none. See text for further explanation.

We should now review some of the trivial mathematical
properties of ϵ^x, and their analogs in biological intuition.
  ϵ^x can be expanded in Taylor series as $1 + x + x^2/2!$
$+ x^3/3! + \ldots$., each of the successive terms being smaller.

(See any elementary calculus text or Smith 1968.) Then $1 - \epsilon^{-\rho\pi r^2}$ equals $\rho\pi r^2 - [(\rho\pi r^2)^2/2! - (\rho\pi r^2)^3/3! + \ldots]$, so that we can say that $1 - \epsilon^{-\rho\pi r^2}$ is always equal to or less than $\rho\pi r^2$. Stated in another way: the projection of the leaves can never be greater than their total surface area; and for a random distribution of leaves, it will generally be less because some of the leaves will overlap.

Of course $1 - \epsilon^{-\rho\pi r^2}$ takes its minimum value, namely 0, when $\rho = 0$; when there are no leaves, they have no projection. $1 - \epsilon^{-\rho\pi r^2}$ can never equal 1 unless ρ is infinite; no matter how numerous the leaves, there are always some gaps in the canopy if leaves are randomly distributed. The only way to cover these gaps is to place new leaves directly under the gaps in the old canopy; these new leaves are no longer distributed independently of the old ones, and the over-all distribution of leaves is therefore spaced rather than random.

ADAPTIVE ADVANTAGES OF MONOLAYER AND MULTILAYER

It is now time to present the two alternative strategies and to begin to discuss their differences. First there is a monolayer, in which, as the name implies, the leaves are spread in a single horizontal layer. Once the leaves reach a certain density in this monolayer, they begin to overlap and the shaded portions are a net loss to the tree. Therefore the optimal horizontal distribution of leaves in the monolayer is spaced rather than random, with new leaves tending to fill the gaps between old ones.

The alternative strategy is a multilayer, in which the leaves are distributed vertically in such a way that leaves whose horizontal projections overlap are at least 70 leaf diameters apart vertically. As long as the average light intensity below these leaves is greater than 20% of full sunlight, new leaves can be added that will photosynthesize at full capacity whether or not they overlap horizontally with leaves above them. Thus the leaves can be distributed independently of each other, i.e., randomly. Since the sun actually moves in a semicircle around these leaves during the course of the day, it is not clear what specific distribution of leaves in several layers would be optimal, and in particular it is not clear that any given dependent distribution would be significantly better than a random one.

We can now compare the two strategies: the monolayer with its projection of 1 or $\rho\pi r^2$, whichever is smaller, and the multilayer with its projection of $1 - \epsilon^{-\rho\pi r^2}$.

The total flux of light that each of these strategies intercepts, and thus the amount that must ultimately be reflected, stored as food, or dissipated as heat, is proportional to the horizontal projection; in fact projection is defined as the proportion of impinging flux intercepted. Therefore the potential heat load of a plant depends on the extent of its projection. For a monolayer, this projection is about $\rho\pi r^2$ or 1, whichever is less; for a multilayer the projection is $1 - \epsilon^{-\rho\pi r^2}$, which is equal to or less than $\rho\pi r^2$. Thus the heat load of a monolayer is at least as heavy as that of a multilayer with the same density of leaves, and for dense foliage it is heavier. Leaves with a high heat load must transpire at a greater rate than leaves that have a low heat load; thus the water required per unit of photosynthesis would be more for a monolayer than for a multilayer. Since water requirement per unit of photosynthesis is an inverse measure of drought resistance, the multilayer should be more drought resistant than the monolayer.

The vertical range over which leaves must be distributed to take advantage of the multilayered strategy is directly dependent on the average diameter of the largest circle that can be inscribed in the leaf. Therefore small leaves or deeply lobed leaves allow greater freedom in the vertical distribution of multilayered foliage. Conversely, the most effective way to assure that spaces between small leaves in a monolayer are covered is to make the leaves larger. Thus the optimal leaf size for the multilayer is smaller than that for the monolayer in the same climate. Smaller leaves add to the drought resistance of the multi-layer, since smaller leaves more effectively dissipate the heat generated from the intercepted radiation. Vogel (1968) has argued that the closer each bit of leaf area is to the edge of the leaf, the more effectively convection currents can carry away accumulated heat. Knoerr and Gay (1965) have presented the relevant theory and have measured the equilibrium temperature of lighted leaves to confirm the effective heat transfer of small leaves and to show that it is indeed related to currents of air, since a wind as light as 2 meters per second cools even large leaves to near the ambient temperature.

Of the total intercepted light flux, only a portion is useable for photosynthesis. The monolayer can cover the available ground area with an equivalent area of leaves, all lit by full sunlight and therefore operating at full photosynthetic rate. The multilayer can expose a larger leaf area, and all leaves will operate at full capacity as long as the lowest leaves are lit by 20% or more of full sunlight. That is, as long as the total projection of leaves is 0.80 or less. Mathematically we want $1 - \epsilon^{-\rho\pi r^2}$ to be less than 0.80; solving for the largest ρ that allows this, we get $\rho = -(\log 0.20)/\pi r^2 = 1.6/\pi r^2$, or at least 1.6 times as much leaf area as there is ground area, all operating at full photosynthetic rate. Thus the maximum growth rate for the multilayer in the open is greater than that of the monolayer.

Since the multilayer holds its growth advantage only when the light below its foliage is greater than 20% of full sunlight, we can ask whether it holds its advantage over the monolayer when both are growing under a canopy that intercepts some of the impinging sunlight. We know that as long as the light below the canopy is greater than 20% of full sunlight, the monolayer can cover the available ground area with leaves, all operating at the peak photosynthetic rate. As long as the multilayer can produce more leaf area than there is ground area and still not reduce the light on its lowest layers to much less than 20% of full sunlight, it can grow faster than the monolayer. Hence for the multilayer to grow faster than the monolayer, the impinging light level times the proportion of this light that penetrates foliage equivalent to a unit of ground area, must be greater than about 20%. Mathematically the impinging light, L_0, times ϵ^{-1} must be greater than 20%. Solving for the minimum L_0 that allows this, we get $L_0 = (0.20)\epsilon = 0.54$. Therefore the multilayer can grow faster than the monolayer only when exposed to between about 54% and 100% of full sunlight. Under conditions much shadier than 54% of full sunlight, the monolayer can grow faster than the multilayer. Thus the monolayer is more shade tolerant than the multilayer. (I here mean absolute shade tolerance rather than the ability of a species of sapling to tolerate the shade cast by a canopy of the same species.)

The shade that each strategy sheds is proportional to its projection. The shade shed by a monolayer is always deeper than that of a multilayer with the same amount of foliage, since fewer leaves of the monolayer overlap each other. Also the maximum shade shed by a monolayer is complete, at least in theory, while no matter how much foliage is distributed in a multilayer, there are still openings.

PREDICTED PATTERNS OF MONOLAYER
AND MULTILAYER

The above properties of the two strategies are direct consequences of the optimal spacing of leaves in either pattern. These and further, indirect consequences are summarized in Table 4.1.

TABLE 4.1. Adaptive advantages of monolayer and multilayer. The numbers represent orders of magnitude, appropriate only to the simplest derivation of this table. A more rigorous formulation (Chapter 5) gives slightly different numbers, but the qualitative conclusions remain the same.

Property	Monolayer	Multilayer
INTRINSIC PROPERTIES		
Photosynthesis in sunlight varies with the:	projection ($\rho \pi r^2 \leq 1$)	leaf area if $\rho \pi r^2 \leq 1.6$
Heat load varies with the:	projection ($\rho \pi r^2 \leq 1$)	projection ($1 - \epsilon^{-\rho \pi r^2}$)
Leaf distribution in layer	uniform	random
Heat load/leaf area	high	low
DIRECT EXTRINSIC PROPERTIES		
Leaf size	large	small or lobed
Drought resistance	low	high
Growth rate in the open	low	high
Absolute shade tolerance	tolerant	intolerant
Shade shed	deep	light
DERIVED EXTRINSIC PROPERTIES		
Height distribution	understory	canopy
Climatological distribution	wet	dry
Successional stage	late	early; also late in xeric areas

Within a fully developed forest there is, of course, a gradient in light intensity from the canopy to the understory, and a corresponding gradient in the saturation deficit, or evaporative power, of the air (Evans 1939, Baumgartner 1967). Thus the multilayered strategy, adapted to high light intensities and to dryness, should prevail in the canopy, but tolerant and drought-sensitive monolayers should predominate in the understory. Multi-layered trees in the canopy face a problem in adaptive development since they must regenerate in the under-story where the monolayer is the appropriate strategy.

Similarly, saplings have different light and moisture environments at different stages of succession (Ross 1954). In an open field they are exposed to full sunlight and to the drying effects of heat and wind, conditions similar to those in the canopy of a fully developed forest. In the later stages of succession, saplings must grow in the under-story. In the early stages of succession the dominant trees will be those that grow fastest to form a canopy. In the full sunlight of an open field, the multilayer grows much faster than the monolayer. As a canopy, however, the multilayer sheds little shade, and it is open to invasion. The monolayered sapling holds a growth advantage over the multilayer when both are invading the shaded under-story. Moreover, when the monolayer reaches the canopy, it sheds a deeper shade and is more resistant to further invasion than the multilayer. Thus succession should start with the invasion of a field by a multilayer and pro-ceed toward a monolayered climax. However, on dry sites a multilayer might persist in the climax due to its drought resistance.

DISCUSSION

This interpretation is sufficient to explain why forest succession starts with the invasion of trees that are intol-erant of shade and proceeds toward more tolerant trees rather than simply starting with the most tolerant species. The theory shows why adaptation of leaf placement for optimal growth in an open field is inconsistent with adap-tation for maximum shade tolerance; that is, why tolerant trees necessarily grow more slowly than intolerant trees in full sunlight (Martin 1959, Grime and Jeffrey 1964). The theory also suggests why late stages of xeric succession have many species that are characteristic of intermediate stages of mesic succession (Jones 1945). Even some primary successions should proceed from multilayer to monolayer as dry, rocky or sandy soils evolve increased capacity for water.

The multilayer should persist into the climax forest in xeric environments and where regular, spotty deaths are caused by drought, seasonal flooding, or shading by vines. In xeric environments the multilayer is adaptive by reason of its low heat load per unit of photosynthesis. Where single, large trees die regularly and often, the climax forest may actually be a patchwork of successional stages. Persistent multilayers continually invade new openings in the forest and race to the canopy where they must also grow quickly to dominate the opening before it is filled with branches of the surrounding canopy trees.

The theory is a generalization of the explanation proposed by Kramer and Decker (1944) for the difference in photosynthesis of pine and broadleafed trees at different light intensities. They suggested that mutual shading of leaves allows photosynthesis of pine seedlings to increase with light intensity even at intensities near full sunlight, while the photosynthesis of pine needles per unit of lit leaf area reaches saturation at the same low light intensities as do broadleafed trees (Kramer and Clark 1947).

The theory also provides a plausible explanation for the gradient in leaf size from large leaves in moist climates to smaller leaves in dry climates (see Daubenmire 1968 for a discussion). Furthermore it suggests additional advantages to the gross morphology of deeply lobed sun leaves and shallow-lobed shade leaves (Vogel 1968), and it explains the developmental strategy of trees like Black Oak that have large, shallow-lobed leaves as saplings, but smaller, deeply lobed leaves as canopy trees (Figure 4.6). Fir, Redwood, and Giant Sequoia leaves show a similar adaptation, having diverse orientation in sun twigs of trees, but being horizontal in shade twigs and seedlings (Büsgen and Münch 1929). Conversely, some early-successional cedars and junipers have leaves in diverse orientations as saplings in open fields, but the leaves are more nearly aligned in a single plane in older trees that are more

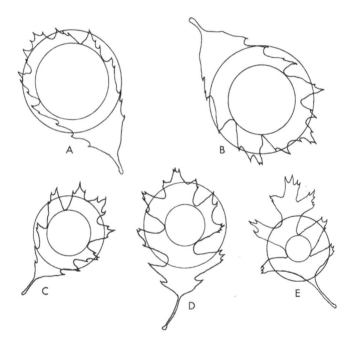

FIGURE 4.6. Leaves of Black Oak.
A. Leaf of a seedling.
B. Leaf from a shaded branch near the ground.
C-E. Leaves from progressively higher on the tree.
   The small circle is the largest circle that can be inscribed in each leaf. The larger circle has the same surface area as the leaf. The relative sizes of the two circles show how well lobing adapts the leaf for a multilayered distribution.

likely to be surrounded by competing species that intercept a significant amount of light (Lyr in Lyr et al. 1967). Some pines show a similarly appropriate ontogenetic pattern, accompanied however by differences in the photosynthetic responses of individual leaves of different ages (Bormann 1958).

The theory suggests that on moist soils the pioneers in succession should have numerous small leaves scattered throughout the tree and oriented in no particular direction, while the climax species should have an outer shell of leaves in a planar orientation. Trees at intermediate seral stages should mix the strategies, perhaps having an outer shell of randomly oriented leaves, or several sparse layers of planar leaves. This interpretation is consistent with most known successions in the north temperate zone, which proceed from pines, junipers, birches, poplars, or larches, through seral stages of firs, maples, gums or oaks, to a climax of beech, hemlock, or spruce.

SUMMARY

This chapter examines the optimal parameters and adaptive advantages of two extreme spatial distributions of leaves: a monolayer, with leaves densely packed in a single layer, and a multilayer, with leaves loosely scattered among several layers.

Direct sunlight is completely blocked by a leaf for a distance of up to 50 or 70 times the minimum diameter of the leaf. Diffuse skylight is significantly blocked for a distance of about one leaf diameter. Therefore the minimum distance between layers of an optimal multilayer is very large in sunny climates, and significantly smaller in cloudy climates. Correspondingly the size of a leaf that is optimal for a fixed distance between layers is very small in a sunny climate, larger in a cloudy climate.

The net photosynthesis of leaves increases with light intensity up to an intensity of about 20% of full sunlight, at which point photosynthesis is near its maximum rate and is unaffected by further increases in light intensity. Thus whenever there is enough foliage to reduce the light at the forest floor to less than 20% of full sunlight, strategies of light interception are important. In particular, the light intensities in the lower layers of a multilayer may be so low that the lower leaves cannot pay for their own respiration. Conversely in the open the multilayer can put out several layers of leaves, to the monolayer's one, without reducing the light on the lowest layers to less than 20% of sunlight. The monolayer, therefore, grows faster in shade than the multilayer, but the multilayer holds an advantage in the open.

The optimal distribution of leaves in a monolayer is a regular spacing of large leaves so that few leaves overlap and there are few gaps between leaves. The leaves of a multilayer can be randomly distributed as long as the leaves are small enough such that those that overlap horizontally are far apart vertically and do not shade each other. The amount of light that penetrates a random distribution of leaves is a negative exponential function of leaf area. Thus the multilayer would need an infinite leaf area to cast complete shade. The light that is intercepted but not used must be dissipated as heat. This heat load is concentrated in a single layer of leaves in the monolayer, but in the multilayer it is spread over several layers. Therefore the multilayer is more resistant to drought than the monolayer. Moreover, the small leaves of the multilayer increase the efficiency with which convection currents can carry away its accumulated heat.

The predictions made by the theory are legion. Leaf size should increase from dry, sunny climates to moist or cloudy climes. Tolerant trees should generally shed deep shade and grow slowly in full sunlight. The adaptation for rapid growth in the open is inconsistent with adaptation for tolerance of shade. Multilayers should dominate the hot, dry, and sunny canopy; but monolayers, the cool, moist, and shady understory. Secondary succession should start with the invasion of an open field by multilayers and proceed to monolayers that invade the understory. Even some primary successions should proceed from multilayers to monolayers as the soil evolves greater capacity for water. Droughts will prevent monolayers from competing effectively with multilayers at late stages of succession, thus causing xeric successions to resemble truncated mesic successions. Chronic disturbances will also bias the late stages in favor of multilayers.

On moist soils the pioneers in succession should have numerous small leaves scattered throughout the tree with no particular orientation. The climax species should have an outer shell of leaves in a planar orientation. Trees at intermediate stages should mix these strategies and should have some degree of developmental flexibility.

All of these predictions are consistent with known qualitative patterns among trees of the north temperate zone. Therefore the theory will be extended in the next chapter to make some roughly quantitative predictions.

CHAPTER FIVE

Photosynthetic Response
of the Strategies

This chapter develops the theory of Chapter 4 in a more rigorous and quantitative form.

Exact expressions for the photosynthesis of a monolayer or a multilayer are derived by combining expressions for the amount of light that penetrates to each layer of a tree with an expression for the rate of photosynthesis of leaves at different light intensities.

The expressions are the basis of a model that makes realistic predictions about changes in the morphology of trees with succession, and rigorously simulates hypothetical situations to analyze growth, reproduction, and invasion. This analysis uncovers some surprises about the roles of productivity and stable age distributions in maintaining the stable specific composition of the climax forest.

LIGHT AND LEAF DISTRIBUTION

A total density of ρ leaves per unit of ground area is uniformly distributed vertically among n layers within which leaves are spaced so that none overlap. The projection of each of these layers is then $\frac{\rho}{n} \pi r^2$. If the layers are far enough apart, in terms of leaf diameter, for them to act as uniform density filters for each other, then a proportion $1 - \frac{\rho \pi r^2}{n}$ of the incident light gets through the first layer; of this proportion, only $1 - \frac{\rho \pi r^2}{n}$ gets through the second layer, and so on down to the nth layer. The proportion of incident light penetrating all n layers is thus

$$\left(1 - \frac{\rho\pi r^2}{n}\right)^n. \tag{5.1}$$

When $n = 1$ the amount of light coming through the tree is $1 - \rho\pi r^2$. This is obviously the same expression as that in the previous chapter for the amount of light coming through the monolayer with no overlapping leaves.

When n approaches infinity, that is when each leaf is essentially the only one in its layer, the amount of light coming through the tree is

$$\lim_{n\to\infty} \left(1 - \frac{\rho\pi r^2}{n}\right)^n = \epsilon^{-\rho\pi r^2}. \tag{5.2}$$

(This is one of the ways to define ϵ^x; see any elementary calculus text.) This is exactly the formula that I have used in the previous chapter for the amount of light penetrating a multilayer with randomly distributed leaves. I have assumed that the layers are far enough apart to act as uniform density filters for each other; this is equivalent to assuming that the layers are independent of each other. I have also assumed that leaves are distributed evenly among the n layers. When n is large enough so that each leaf is the only one in its layer, these assumptions imply that individual leaves are distributed independently and with uniform density, i.e., randomly.

Thus the extremes of a continuum measured by n are the mathematically convenient cases that I have already called monolayer and multilayer.

PHOTOSYNTHESIS OF INDIVIDUAL
LEAVES

Since we can calculate how much light penetrates to each level of a specified distribution of leaves, we must now find the photosynthetic rate of a leaf exposed to a given amount of light.

The physiological relation between light and photosynthesis is so complex that no simple model can accurately reflect it. However, all that we need is an empirical relation. We can therefore start by considering a leaf to be a simple machine that magically converts light into some useful product. We can then develop a reasonable theoretical relation between photosynthesis and light intensity by using the argument that supports the Michaelis-Menten equation of enzyme kinetics (White et al. 1968).

Assume that light is trapped by some mechanism, and that the energized trap then decays to an unenergized trap and an energized product that is ultimately incorporated into sugar:

$$\text{Trap } (T) + \text{Light } (L) \underset{k_2}{\overset{k_1}{\rightleftharpoons}}$$

$$\text{Energized trap } (TL) \xrightarrow{k_3} T + \ldots \text{Sugar.}$$

The rate of each reaction depends on the concentration of each reactant [represented by square brackets] and the characteristic rate constant k_i of the reaction. $[T]$ represents the concentration of all traps, energized or not. The rate at which TL is formed is $k_1([T] - [TL])[L]$, since it depends on the number of meetings between the unenergized traps and light. The rate at which TL is disappearing is $k_3[TL] + k_2[TL]$, some of it decaying to unenergized trap and product, while some of it loses its light before it can be fixed as a useful product. Thus the rate of change of the amount of energized trap with time (t) is:

$$d[TL]/dt = k_1([T] - [TL])[L] - (k_2 + k_3)[TL].$$

At equilibrium the amount of TL is not changing; hence by setting $d[TL]/dt = 0$, and shifting things around a little, we get:

$$([T] - [TL])[L]/[TL] = (k_2 + k_3)/k_1 = k,$$

where k is just a new constant, a sort of "binding constant" that measures the effectiveness of the trap in getting and processing photons. The velocity of the reaction, which is the rate of total photosynthesis that we want, is $P_t = k_3[TL]$. The maximum possible rate of the reaction would occur when there is so much light that all of the traps are present in the energized form, when all the T is in TL; the saturated rate of photosynthesis is then $P_{max} = k_3[T]$. If we now replace $[T]$ and $[TL]$ by their respective values P_{max}/k_3 and P_t/k_3, and shift around the above expression for k a little, we get:

$$P_t = P_{max}[L]/([L] + k).$$

Since the concentration of light is measured in units of intensity, we can just use light intensity, L, in place of $[L]$. The net photosynthesis P is then the total photosynthesis P_t minus respiration R:

$$P = \frac{P_{max}L}{L + k} - R. \tag{5.3}$$

From this equation we can calculate the light compensation point (C), the light intensity at which photosynthesis just balances respiration.

$$\frac{P_{max}C}{C + k} = R$$

$$C = Rk/(P_{max} - R). \tag{5.4}$$

Of course the physiological relation between light and photosynthesis is much more complex than this model. Many other factors influence the rate of photosynthesis, for example: carbon dioxide, water supply, humidity, chlorophyll recruitment, temperature, and such. I need only consider those factors that vary when the incident light changes, but most of the named factors do. In particular, the parameters R, k_1, k_2, and k_3 will certainly be different at different temperatures, and temperature will in turn vary with changes in the intensity of incident radiant energy. Heath (1969) discusses such details of the mechanism of photosynthesis, and he rightly criticizes the physiological use of simple models that distort the mechanism. However, the modified Michaelis-Menten equation provides a simple empirical representation of the effect of light intensity on photosynthesis. Monteith (1963) has developed the same equation, using a similar argument, and found that it accurately describes the actual photosynthetic responses of several species of herbaceous plants. Figure 5.1 shows that the dimensionless form of Equation 5.5 approximates empirical data from trees closely enough that the theory based on it should not be far wrong. I have plotted the proportion of maximum photosynthesis per unit of lit leaf area against the proportion of full sunlight, and a curve of the form:

$$(P + R)/P_{max} = L/(L + k). \tag{5.5}$$

The parameters are easily fitted. R is the measured respiration rate of the plant in the dark, expressed as a proportion of the peak photosynthesis. The parameter k can be estimated either from parameters of a least squares linear regression of $P_{max}/(P + R)$ on $1/L$, or by solving $P_o + R = L_o/(L_o + k)$ for k using one of the first empirical points for the values of L_o and P_o. The latter procedure gives the better fit to the critical data at low light intensities.

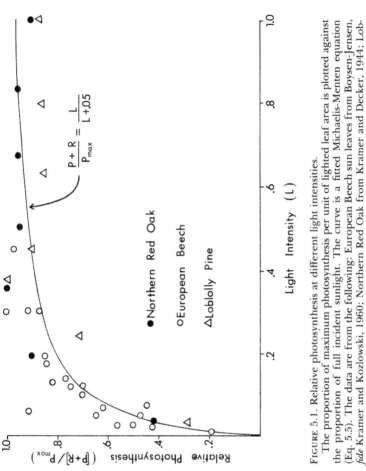

$$\frac{P + R}{P_{max}} = \frac{L}{L + .05}$$

●Northern Red Oak

○European Beech

△Loblolly Pine

Light Intensity (L)

Relative Photosynthesis ([P+R]/P$_{max}$)

FIGURE 5.1. Relative photosynthesis at different light intensities. The proportion of maximum photosynthesis per unit of lighted leaf area is plotted against the proportion of full incident sunlight. The curve is a fitted Michaelis-Menten equation (Eq. 5.5). The data are from the following: European Beech sun leaves from Boysen-Jensen, *fide* Kramer and Kozlowski, 1960; Northern Red Oak from Kramer and Decker, 1944; Loblolly Pine from Kramer and Clark, 1947.

We can take 0.05 as a representative value for the parameter k. I shall use 0.25, the value for Northern Red Oak, for R. Thence $C = 0.015$, an appropriate order of magnitude for the compensation intensity of several species of shade tolerant trees (Baker 1950). More data to confirm these values would be welcome, but exact values are not critical to the theory.

RATIONALE FOR OPTIMAL PHOTOSYNTHESIS IN A MULTILAYER

We can now calculate the net photosynthesis for a tree with leaves randomly distributed in a horizontal plane and with an arbitrary vertical distribution of densities.

Let the density of foliage at a given height h be $F(h)$. To simplify the math, we will measure h as the distance down from the top of the canopy. The light intensity at a given height is

$$L_h = L_0 \epsilon^{-\int_0^h F(h)dh},$$

and the height at which light reaches the compensation point, h_c, is defined by

$$C = L_0 \epsilon^{-\int_0^{h_c} F(h)dh}.$$

The net photosynthesis above a unit of ground area is the integral of the density of leaves at each height times the amount of photosynthesis per unit area of leaves that are exposed to as much light as gets through the leaves above that height. Since we are looking for an optimal strategy, we integrate from the top of the canopy down to the height where light intensity is so reduced that photosynthesis just balances respiration, the light compensation point. The net photosynthesis above a unit of ground area is then:

70

$$\int_0^{h_c} F(h) \left[\frac{P_{max}L_0\epsilon^{-\int_0^h F(h)dh}}{L_0\epsilon^{-\int_0^h F(h)dh} + k} - R \right] dh.$$

This integral is solved by splitting off the R part and integrating it separately; the rest is integrated by substituting the variable

$$Z = L_0\epsilon^{-\int_0^h F(h)dh} + k.$$

Then

$$dZ = -L_0F(h)\epsilon^{-\int_0^h F(h)dh} dh;$$

and the limits of integration are:

$$Z_{h_c} = L_0\epsilon^{-\int_0^{h_c} F(h)dh} + k = C + k$$

$$Z_0 = L_0\epsilon^{-\int_0^0 F(h)dh} + k = L_0 + k.$$

Thus:

Photosynthesis of optimal multilayer per ground area $= -P_{max} \int_{L_0+k}^{C+k} \frac{dZ}{Z} - R \int_0^{h_c} F(h)dh$

$$= P_{max} \log \frac{L_0 + k}{C + k} - R \log \frac{L_0}{C}. \qquad (5.6)$$

It is significant that the actual height distribution of foliage density does not appear in the final expression for net photosynthesis; the only parameters are constants from the equation for the effect of light on photosynthesis, P_{max}, the rate of light-saturated photosynthesis, k, the binding constant that measures the effectiveness of leaves in trapping and processing photons, and C, the light intensity at which photosynthesis just balances respiration.

We reach the conclusion that the exact height distribution of randomly placed leaves is irrelevant to their efficiency as long as they are spread over a sufficient vertical range to minimize mutual shading by leaves that overlap horizontally. This result should have been obvious a priori (although I did not realize it until I had gone through the above machinations), since height is just a convenient dummy variable over which we are really integrating by leaf area.

PHOTOSYNTHESIS IN A MULTILAYER

The original expression for the photosynthesis of a multilayer is much more tractable if horizontal leaf area (A) is actually used as the variable of integration. If we integrate from the most brightly lit leaves to the least, then the amount of light reaching a given leaf is a negative exponential function of the total leaf area between the given leaf and the sky. The photosynthesis per unit of ground area for a multilayer with a density of ρ leaves per unit of ground area, lit by incident light of intensity L_0 is then:

$$\text{Photosynthesis per ground area} = \int_0^{\rho\pi r^2} \frac{P_{max}L_0\epsilon^{-A}}{L_0\epsilon^{-A} + k}\, dA - \rho\pi r^2 R$$

$$= P_{max} \log \frac{L_0 + k}{L_0\epsilon^{-\rho\pi r^2} + k} - \rho\pi r^2 R. \quad (5.7)$$

The optimal density of leaves is that which just reduces the light intensity below the tree to the compensation point (C); as long as there is more light, additional leaves would add more by photosynthesis than they would subtract by respiration, but once the light reaches the compensation point, additional leaves respire more than they photosynthesize. The optimal density of leaves is therefore defined by $L_0\epsilon^{-\rho\pi r^2} = C$. Note that L_0 enters the expression for the optimal density of leaves; thus the optimal strategy requires a different density of leaves for each different intensity of incident light. When we replace $L_o \epsilon^{-\rho \pi r^2}$ by C in the above formula we get:

Photosynthesis per ground area for multilayer with optimal leaf density
$$= P_{max} \log \frac{L_o + k}{C + k} - R \log \frac{L_o}{C}.$$

(5.8)

It is comforting to note that this is the same as Equation 5.6 of the previous section.

Again, even if the details of the photosynthetic response of a whole tree do not conform to my simplifying assumptions, the expression should be empirically accurate for a tree with several layers of leaves, for the expression is based on an exponential extinction of light by leaves. Monsi and Saeki (1953) have shown that light intensity within a stand obeys Beer's law; that is, it falls off at an exponential rate dependent on leaf area:

$$L(h) = L_o \epsilon^{-Ka(h)},$$

where K is a fitted constant and $a(h)$ is the amount of leaf area per unit of ground area above height h. Miller (1969) has shown that an expression very like $(1 - \rho \pi r^2/n)^n$ gives an even better fit to the empirical relation between leaf area and light penetration. Furthermore, Monteith (1965) has used a similar formula, together with the modified Michaelis-Menten equation, to obtain an expression for the net photosynthesis of plants with several layers of leaves. He found that his expression predicted accurately the net photosynthesis of several species of crop plants.

PHOTOSYNTHESIS IN A MONOLAYER

The monolayer exposes a single layer to light of intensity L_o; thus its net photosynthesis per unit of ground area is simply:

$$\begin{aligned}
\text{Photosynthesis} \atop \text{per ground area} &= \rho\pi r^2 \frac{P_{\max}L_o}{L_o + k} - \rho\pi r^2 R \qquad \text{for } \rho \leqslant \frac{1}{\pi r^2}\\[2ex]
&= \frac{P_{\max}L_o}{L_o + k} - \rho\pi r^2 R \qquad \text{for } \rho > \frac{1}{\pi r^2} \quad (5.9)
\end{aligned}$$

As long as L_o is above the compensation point the optimal density of leaves for photosynthesis is a complete layer with no holes, or $\rho\pi r^2 = 1$. Of course if L_o is below the compensation point, then none of the leaves can have a positive net photosynthesis. Thus the best that a monolayer can do is to put out a complete layer of leaves:

$$\text{Photosynthesis per ground} \atop \text{area for monolayer with} \atop \text{no gaps between leaves} = \frac{P_{\max}L_o}{L_o + k} - R. \quad (5.10)$$

MULTILAYER VERSUS MONOLAYER

After assuming common parameters for the leaves of a multilayer and a monolayer ($k = 0.05$; $C = 0.015$; and $R/P_{\max} = 0.25$), we can plot the optimal photosynthesis for either strategy at different intensities of incident light (Figure 5.2). The curves cross. In an open field, the multilayer photosynthesizes faster than the monolayer; in the shade the monolayer holds the advantage. Furthermore, the light intensity at which the curves cross is an appropriate order of magnitude for the light on a forest floor at an intermediate stage of succession (Table 3.1 in Chapter 3). Thus the qualitative predictions of Table 4.1 in Chap-

74

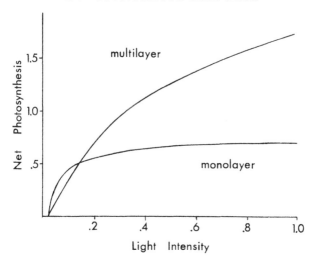

FIGURE 5.2. Effect of light on photosynthesis of multilayer and monolayer.
   Light intensity is measured as proportion of full sunlight. The equations for photosynthesis by optimal multilayer (Eq. 5.8) and optimal monolayer (Eq. 5.10) use the parameters: $P_{max} = 1.0, R = 0.25; k = 0.05$. Thus $C = 0.015$.

ter 4 are consistent with a more exact model of the effect of leaf distribution on photosynthesis. Multilayers should dominate the sunny canopy, but monolayers, the shaded understory. Succession should start with the invasion of an open field by multilayers and proceed to monolayers that invade the understory.

From Figure 5.2 we have the further prediction that changes in the specific composition of the forest will be much more rapid during the early stages of succession than during the later stages, since the advantage of the multilayer over the monolayer and the growth rates of both at high light intensities are much greater than the advantage of the monolayer and both growth rates at low light levels.

75

The extremes of a monolayer with no gaps and no over-lapping leaves and a multilayer with an infinite number of layers are somewhat unrealistic. However it is possible to calculate the photosynthesis of a tree with an arbitrary number of layers and to compare it with the extreme monolayer and multilayer. The rationale is the same as that for calculating the photosynthesis of a multilayer, ex-cept that a finite expression is used for the amount of light reaching each layer, and photosynthesis is summed over the n layers rather than being integrated over the $\rho \pi r^2$ leaf area.

Photosynthesis
per ground area

$$= \sum_{i=1}^{n} \left[\frac{P_{max} L_o \left(1 - \dfrac{\rho \pi r^2}{n}\right)^{i-1}}{L_o \left(1 - \dfrac{\rho \pi r^2}{n}\right)^{i-1} + k} \left(\frac{\rho \pi r^2}{n}\right) \right] - \rho \pi r^2 R. \quad (5.11)$$

Again this will be optimal if the tree has just enough leaves to reduce light at the lowest layers to the compensation point C. That is if:

$$L_o \left(1 - \frac{\rho \pi r^2}{n}\right)^{n} = C \qquad (5.12)$$

whence:

$$\left(1 - \frac{\rho \pi r^2}{n}\right) = \sqrt[n]{\frac{C}{L_o}}$$

$$\frac{\rho \pi r^2}{n} = 1 - \left(\frac{C}{L_o}\right)^{\frac{1}{n}}$$

$$\rho \pi r^2 = n - n \left(\frac{C}{L_o}\right)^{\frac{1}{n}}.$$

Substituting these into the above expression for photosynthesis, we get:

Photosynthesis per ground
area for an n-layer with
optimal leaf density

$$= \left[1 - \left(\frac{C}{L_0}\right)^{\frac{1}{n}}\right] \left[P_{max} \sum_{i=1}^{n} \left(\frac{L_0 \left(\frac{C}{L_0}\right)^{\frac{i-1}{n}}}{L_0 \left(\frac{C}{L_0}\right)^{\frac{i-1}{n}} + k}\right) - nR\right]. \quad (5.13)$$

This intractable expression is plotted in Figure 5.3, using the parameters $k = 0.05$ and $R/P_{max} = 0.25$. From the figure it is obvious that the advantage to be gained by multiple layers is appreciable with as few as two layers, and that 10 layers is a reasonable approximation to an infinite number.

The following account of the relations among the foregoing equations is intended solely for those who cannot yet accept Figure 5.3 on the basis of logic, intuition, or faith. The most general equation is 5.11; it can be converted directly to 5.9 by setting $n = 1$; it can be converted to its limiting form, 5.7, by letting n approach infinity, that is by applying 5.2. Similarly 5.13 can be converted to 5.6 or 5.8 by letting n approach infinity. Setting $n = 1$ in 5.13 results in an expression that is $(1 - C/L_0)$ times Expression 5.10, because the condition set by 5.12 is not quite optimal for $n = 1$ when L_0 is just slightly larger than C. The condition of 5.12 requires that a tree set out only enough leaves to reduce light to the compensation point below its lowest layer. If a tree that uses this strategy has very few layers, it could increase its net photosynthesis by filling the gaps in its lowest layer, thus achieving the optimal leaf distribu-

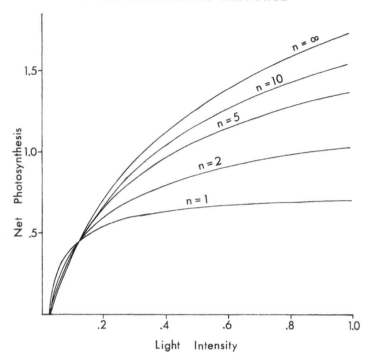

FIGURE 5.3. Effect of light on photosynthesis by a tree with n layers. Light is measured as proportion of full sunlight. The optimal leaf density is assumed at each light intensity, in accordance with Equation 5.10 for $n = 1$, Equation 5.8 for $n = \infty$, and Equation 5.13 for $n = 2, 5, 10$. The parameters are: $P_{max} = 1.0$, $R = 0.25$; $k = 0.05$.

tion for a set number of layers. However, the increase is virtually undetectable except for a slight but noticeable increase for a monolayer near the compensation intensity of light. Equations 5.8 and 5.13 are valid only for $L_o > C$; the definitions of optimum leaf density would allow negative values of ρ for $L_o < C$. Negative leaves cannot be admitted, even in the theory.

RELAXING THE ASSUMPTIONS

For ease in the analysis, I have assumed that all leaves are horizontal and that all light comes from directly above. If all light did come from directly above, the optimal multilayer should droop leaves in its upper layers to intercept the light flux at such an angle that the light intensity on the leaf's surface were near the level at which the photosynthetic mechanism saturates. Drooping leaves in the upper layers would then have a lower heat load per unit of photosynthesis than horizontal leaves, and the upper layers would let through more light, allowing one or more paying layers to be added at the bottom of the tree. However, the sun moves across the sky through half of a circle; so the leaves in the upper layers of a multilayer must continually adjust their pendent angle over a wide range during the course of a day, if they would take full advantage of the available light. Although it is impossible to design a single, fixed strategy that keeps all leaves of a multilayer uniformly illuminated at all times, drooping high leaves and horizontal low leaves approaches an optimal strategy because the high leaves receive relatively more light from the sun near the horizon than low leaves. Verhagen, Wilson, and Britten (1963) have postulated an ideal multilayer in which the exponential constant of light extinction by leaf area increases with depth in the canopy. I have shown that the height distribution per se of horizontal leaf area is irrelevant to photosynthesis, as long as the layers are far enough apart that leaves in one layer do not eclipse leaves in the next layer. Approximations to Verhagen's ideal multilayer can be made, however, by drooping the upper layers and holding the lower layers horizontal. Thus allowing leaves to droop adds more paying leaf area to the multilayer at a lower heat load than my theory suggests, but since it only enhances the advantages of multilayer over monolayer in the open and does not affect the strategies in the shade, it enhances the patterns that I have predicted.

The fact that the sun moves across the sky rather than staying at the zenith invalidates exact quantitative prediction of productivity from my theory, but the qualitative conclusions about differences between monolayers and multilayers remain unchanged. The movement of the sun, by itself, should only introduce a fluctuating correction factor that is the same for monolayer as for multilayer. The effect of the temporal variation of light on the temporal average of photosynthesis is more complex (Bormann 1956). For example, a leaf can spend $1 - \dfrac{R}{P_{max}}$ of its time in total darkness as long as it spends the rest of its time in enough light to photosynthesize at a rate P_{max}. Consequently, leaves that are completely shaded at a given time will be lighted at other times, perhaps often enough to pay for themselves. This difficulty should not affect the predictions that are based on strategies of leaf placement because when these leaves are lit, others will be shaded. However, more paying leaf area will accompany the multilayered strategies than the present theory allows. In particular much of the overlap between leaves in a single branch will be photosynthetically useful.

I have also ignored the known physiological differences between sun leaves and shade leaves. In many species, leaves that grow in full sunlight have a higher compensation point than shade grown leaves; they also approach saturation at higher light intensities and photosynthesize at a higher rate in full sunlight. Therefore shade leaves operate more efficiently in the shade than sun leaves, but sun leaves trap more energy than shade leaves in the open. If leaves in each layer of a tree were optimally adapted to the average light at that level, then the photosynthesis of an optimal multilayer in the open could be higher than it could if all leaves in the tree had the same physiological response to light. In deep shade, an optimal multilayer must have very few layers lest the light on the lowest layer be reduced below the compensation point. With few layers there is only a small range of light intensities within the foliage, and there is little to be gained from variations in the physiological response of leaves to different light intensities. Thus sun and shade leaves increase the differences between monolayer and multilayer but leave the predictions from the theory intact. Incidentally, a microscopic version of the theory of leaf distribution may help to interpret the physiological differences between sun and shade leaves. One of the morphological differences between sun and shade leaves is the over-all distribution of chloroplasts within the leaf. In shade leaves the chloroplasts are concentrated into a thin layer near the upper surface of the leaf; in sun leaves the chloroplasts are more loosely distributed through several layers of palisade cells (Büsgen and Münch 1929). Figure 5.4 shows this for sun and shade leaves of American Beech. Jackson (1967) has examined leaf anatomy in 21 species of deciduous trees, finding a single palisade layer in all tolerant species and in nearly all shade leaves; the sun leaves of most intermediate and intolerant species have 2 or 3 palisade layers. Thus chloroplasts show "monolayer" and multilayered distributions within leaves adapted respectively to low and high light intensities.

All of the above effects could be incorporated into the theory as variations in the parameters: R, the mean respiration of leaves, k, the Michaelis-Menten constant of photosynthesis, and P_{max}, the light-saturated rate of photosynthesis. The effects of temperature, chlorophyll recruitment, and nutrient concentrations could be similarly accommodated. The parameters R, k, and P_{max} doubtless have adaptive patterns of their own which deserve study.

81

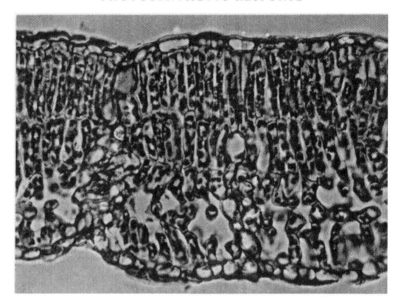

A

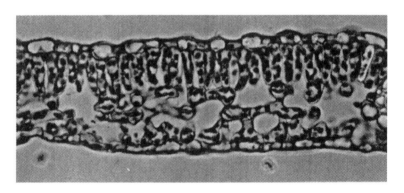

B

FIGURE 5.4. Leaves of American Beech in cross-section.
A. Sun leaf.   B. Shade leaf.
   The dark blobs are chloroplasts, which are distributed in many
more layers in the sun leaf than in the shade leaf.

PHOTOSYNTHETIC RESPONSE

As this important problem is untouched, I can only use sample values of R, k, and P_{max}. However, Figure 5.5 shows that large variations in these parameters do not

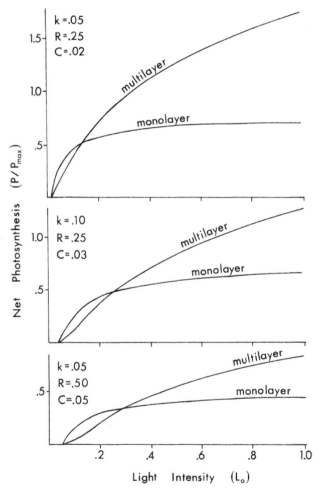

FIGURE 5.5. Photosynthesis of optimal multilayer (Eq. 5.8) and monolayer (Eq. 5.10) for various· values of the photosynthetic parameters of individual leaves.

$P_{max} = 1.0$ in all cases.

affect the relative behavior of curves for photosynthesis of monolayer and multilayer at different light intensities. The larger R and k are, the greater the differences between monolayer and multilayer, the higher the compensation point, and the higher the light intensity at which the curves cross.

Thus most of the difficulties that limit exact quantitative predictions appear to increase the predicted differences between monolayer and multilayer. The qualitative form of the theory seems, therefore, to be robust.

GROWTH, REPRODUCTION, INVASION, AND STABILITY

What I have plotted for the multilayer is the optimal strategy at each light level, which reduces the light below the tree to the level at which photosynthesis just balances respiration. Therefore an ideal stand of an optimal multilayer could not be invaded by a monolayer, but neither could saplings of the multilayer grow in their parents' shade. If the multilayer were not optimal, but shed shade only to the point where a multilayered sapling still holds an advantage over a monolayer (i.e., more light than the point where the optimal curves cross), its curve would lie wholly below that of the optimal multilayer. Although this sub-optimal multilayer would be able to reproduce in its own shade, it would lose in competition with an optimal multilayer, both in the open and in its own shade. Even a slightly denser multilayer would outcompete the sub-optimal multilayer, as long as its foliage is not denser than that of an optimal multilayer.

At any point below a dense multilayer where the density of leaves falls short of the optimal density, a monolayer can invade and grow faster than saplings of the multilayer. The most efficient monolayered invader will shed complete shade, and thus cannot grow in its own shade.

Two important generalizations emerge from the theory. First, it is impossible to design a single morphological strategy that allows a tree both to regenerate in its own shade and to prevent invasion by a species with a more shade-adapted morphology. Second, monolayered trees in the later stages of succession cannot reproduce in their own shade, unless their saplings have some source of energy other than their own leaves, like roots that are grafted to those of older trees.

Reproduction should be more apparent in the early stages of succession than in later stages. Therefore populations of trees in the early stages may have age structures that are more nearly stable than those of later stages. In fact the later stages may characteristically have senile age distributions. The stability of the climax forest, if it exists, is not a stability of age structure but a stability of species composition, which depends on senile trees occupying a large enough area that they can reproduce in the spaces where a conspecific dies, before the seeds of an early successional species arrive.

Furthermore, when the monolayer reaches the canopy after outcompeting multilayers in the understory, only a single layer of leaves is exposed to full sunlight, and the productivity of each unit of ground area is lower than at earlier stages of succession when the ground was occupied by multilayers exposed to full sunlight. Thus the theory predicts that the climax forest should have a senile age structure and a lower productivity than the stages that preceded it. As predicted, many virgin forests are indeed "overmature" (Hellmers 1964, Whittaker 1966); this fact has been an embarrassment to previous theories of forest succession.

The theory can rigorously account for at least three stages in forest succession: 1. Windblown seeds of sparse multilayers adapted to wide dispersion and rapid growth in areas of low root-shoot competition, dry soil, and a high heat load; 2. Heavy-seeded multilayers with denser leaves, which grow faster and face higher water requirements with a more efficient root system; 3. Once light is reduced below the point where the photosynthetic curves of optimal multilayer and monolayer cross, a monolayer. The theory will admit more stages, and perhaps would demand more if more were known about naturally occurring variations in the parameters that govern the relation between light and photosynthesis in individual leaves.

SUMMARY

Exact expressions for the photosynthesis of a monolayer or a multilayer can be derived by combining expressions for the amount of light that penetrates to each layer of a tree with an expression for the rate of photosynthesis of leaves at different light intensities.

A single expression is derived for the light that passes through a tree with foliage distributed evenly among n layers. When $n = 1$ this expression gives the amount of light that penetrates a monolayer as a linear function of leaf density. When $n = \infty$, that is when there are so many layers that each leaf is essentially the only one in its layer, the same expression gives the amount of light that penetrates a multilayer as a negative exponential function of leaf density. Thus the extremes of a single expression represent the leaf distributions of the preceding chapter: a uniformly spaced monolayer and a random multilayer.

An expression is derived for photosynthesis at different light intensities by an analogy with enzyme kinetics. The leaf is viewed as an enzyme that temporarily and reversibly binds its substrate, light, before converting it irreversibly to a sugar. The Michaelis-Menten equation then describes the relation between light and photosynthesis.

Even though the analogy is physiologically unrealistic, the equation is empirically accurate.

The expressions for light at different levels and for photosynthesis at different light intensities are combined to express the photosynthetic rates of monolayers, multilayers, and trees with any intermediate number of layers. When photosynthesis is plotted against light intensity, the curves of monolayer and multilayer cross; the monolayer grows faster at low light intensities, but the multilayer grows faster at high light intensities. Two layers of leaves are enough to make a tree behave somewhat like a multilayer rather than a monolayer, and 10 layers of leaves are a very good approximation to the infinite multilayer.

The height distribution of horizontal leaf area in a multilayer is irrelevant to its photosynthesis as long as the layers are far enough apart so that no leaf completely blocks sunlight from parts of leaves in the next layer. If upper leaves droop and lower leaves are horizontal, then the advantage of the multilayer at high light intensities is increased. The movement of the sun adds the possibility of further low layers where leaves are shaded most of the time, but fully lit often enough for photosynthesis to balance respiration. Multilayers can derive still further benefits from appropriate distribution of leaves adapted to sun and shade. At a microscopic level the chloroplasts are distributed in multilayered fashion in sun leaves, but in monolayered fashion in shade leaves. All of these variations could be accommodated in the expressions for photosynthesis of monolayer and multilayer, but since they only increase the quantitative response of the multilayer at high light intensities and leave both strategies unchanged at low light intensities, they do not affect qualitative predictions about differences between monolayer and multilayer.

The theory has surprising predictive power for an expression with only three independent parameters. Succession should proceed from multilayers to monolayers, and changes in the specific composition of a stand should be faster in the early than in the late stages. It is impossible to design a single strategy that allows a tree both to re-generate continuously in its own shade and to prevent in-vasion by a more shade-adapted morphology. The most efficient monolayer ultimately sheds complete shade, effectively blocking invasion but also preventing its own reproduction. Hence the climax forest should have a senile age distribution rather than a stable one. Moreover, when the monolayer grows from the understory to the canopy, a grossly multilayered forest is replaced by a monolayer, and the productivity of the forest declines. Thus the theory predicts that overmaturity should be the usual condition of virgin and climax forests.

CHAPTER SIX

Measurement of Actual Strategies

The predictions of the previous chapter are consistent with observations of trees that are different enough to be intuitively classified as monolayer or multilayer. However, more accurate tests demand an objective measure of just how many layers a given tree has. A simple extension of the theory provides such a measure. The predictions of Chapters 4 and 5 are confirmed by data from forests in New Jersey, California, and Costa Rica. The measurements also uncover some early successional multilayers that persist into the climax forest by virtue of the deep shade that they shed, as well as late successional monolayers that remain as a permanent understory in some multilayered climax forests.

THEORY

Recall from Chapter 5 that when ρ leaves are uniformly distributed among n layers the amount of light getting through each layer is $1 - \rho\pi r^2/n$, and the amount of light getting through the whole tree is $(1 - \rho\pi r^2/n)^n$. The monolayer and the multilayer are the extremes of a continuum measured by n, the effective number of layers, which can be measured in the following way. A single branch from a tree is assumed to represent a single layer within which the overlap between leaves is heavily shaded. The projection of leaves in this one layer is measured by the methods of Chapter 2; call it p. Then measure the total proportion of light penetrating the tree to the understory; call it u. If we assume that the branches form layers that are equally dense and independently distributed, then $(1 - p)^n = u$. Solving this for n gives

$$n = \log{(u)}/\log{(1 - p)}. \qquad (6.1)$$

If branches are not uniformly and independently distributed, n can still be calculated as above, but it is no longer interpreted as the actual number of branch-layers. Rather n is interpreted as the effective number of layers, and it is still appropriate for theories involving the number of equally dense layers.

What constitutes a branch is not as much of a difficulty as it would seem at first. The branches of most trees are almost planar within an error of 10 or so leaf diameters. If the leaves are very large, more than one branch might be taken to represent a single layer within which overlapping leaves are effectively shaded; if the leaves are tiny and needlelike, only part of a branch should be taken to represent a single layer, since some leaves that overlap within the branch may be far enough apart such that they do not shade each other heavily. These difficulties do not bias tests of the predictions. Most of the trees with large leaves are monolayers and their number of layers would be slightly overestimated; most of the trees with tiny leaves are multilayers and their number of layers would be slightly underestimated. The only trees that I cannot measure objectively are those with trailing pendant or erect bottle-brush branches: elms, willows, cottonwoods, young ashes, that is most temperate river-bottom species, and most western pines. Unfortunately, all of these trees are some sort of multilayer, and I should like to know their exact numbers of layers.

RESULTS

The numbers of layers in plants at different levels in an oak-hickory forest are found in Table 6.1. There are clearly many layers in the canopy and progressively fewer at lower levels, to a monolayer for the ground cover. This pattern conforms entirely to the predictions of Chapters 4 and 5.

TABLE 6.1. Layers of foliage at different heights in an oak-hickory forest. The number of layers in this and following tables was calculated from Equation 6.1, with measurements of projections made using the methods of Chapter 2.

Height	Species	Mean number of layers ± its standard error	
Canopy	oak-hickory	2.7	±0.2
Understory	Flowering Dogwood	1.4	±0.1
Shrub	*Viburnum acerifolium*	1.1	±0.1
Ground cover	*Podophyllum peltatum*	1.0	

Table 6.2 lists the number of layers of various species of trees that are characteristic of different stages of succession. In this table the percentage of light getting through the tree is less than the values given for the same species in Chapter 3. The measurement appropriate for calculating the number of layers is the amount of light getting through the most densely foliated parts of the tree, rather than the average amount of light getting through the vertical projection of the tree, which includes areas where a single branch sticks out, or even open spots between branches. The amount of light getting through Sassafras trees in the open (Table 6.2) is far less than that getting through Sassafras trees in the forest (Table 3.1 and Figure 3.9 in Chapter 3). The last column in Table 6.2 is the number of layers multiplied by the projection of a single branch. This number is

TABLE 6.2. Layers of foliage in trees of different successional stages. Number of measurements is followed by mean % of skylight penetrating branch and tree, mean number of layers ± its standard error, and effective leaf area index. The number of layers is the mean number of layers for measurements on several trees, not the ratio of the logarithms of the mean proportions of light through several branches and several trees. The number of layers has a much lower variance within each species than either of the other measurements.

Species	# Meas-ured	% Light/ branch	% Light/ tree	# Layers ± SE	Leaf A/gnd. area
EARLY SUCCESSION					
Gray Birch	10	44	3.6	4.3 ±0.4	2.4
Bigtooth Aspen	6	45	6.9	3.8 ±0.5	2.1
White Pine	13	25	0.8	3.8 ±0.4	2.9
Sassafras	3	14	0.8	2.7 ±0.7	2.4
MID-SUCCESSION on moist soil					
ash	10	26	3.0	2.7 ±0.2	2.0
Blackgum	7	15	1.4	2.6 ±0.5	2.2
Red Maple	21	20	1.8	2.7 ±0.2	2.2
Tuliptree	6	17	2.3	2.2 ±0.2	1.8
MID-SUCCESSION (late on dry soil)					
red oak	19	23	2.6	2.7 ±0.2	2.1
Shagbark Hickory	12	18	1.4	2.7 ±0.2	2.2
Flowering Dogwood	13	5	2.1	1.4 ±0.1	1.3
LATE SUCCESSION					
Sugar Maple	8	9	1.2	1.9 ±0.1	1.7
American Beech	16	6	1.5	1.5 ±0.1	1.4
Eastern Hemlock	13	8	2.1	1.6 ±0.1	1.4

in some sense an effective leaf area index, discounting leaves that overlap closely within a branch and effectively shade each other, but also discounting the additional photosynthetic surface of dangling leaves high in the canopy.

The pattern of layering through succession conforms to the predictions made in Chapters 4 and 5. The pioneers are multilayered; climax species are almost monolayered. Mid-successional species are intermediate. The climax on

92

dry soils has more layers than the climax on moist soils; it has roughly the same number of layers as mid-succession on moist soils.

The other numbers in Table 6.2 provide further insights. The amount of light getting through the densest foliage of most mid- and late-successional trees is near or below the compensation point. However some pioneers of open fields (Gray Birch and Bigtooth Aspen) let through enough light to allow another complete layer of leaves, which in addition to allowing increased photosynthesis would also slow the invasion of other species. Perhaps these trees are compromising optimal light interception to take advantage of the lowered heat load and low water requirement of a sparse multilayer. Both features of the sparse multilayer are adaptive to the drying environment of a field and the low water capacity of early-successional soils.

The effective leaf area of early and mid-successional broad-leaved trees is about the same. The conifer has a higher leaf area index, which may be the cause of its higher rate of wood production as a plantation tree. The late-successional species have lower effective leaf area indices. Their high efficiency in a shady environment and their own deep shade, factors to which they owe their late-successional status, have compromised their strategy for optimal light utilization in the open. Here is evidence that when succession is driven by shade tolerance and shade shed the productivity of the climax forest is lower than that of earlier stages.

White Pine has many layers, has a high effective leaf area index, and sheds very deep shade. Correspondingly, it may form either an early stage in succession or a highly persistent stage. White Pine is a pioneer in old fields of the northeastern United States; it is also found as scattered trees in virgin forest and as dense and very old stands, but it is never found invading an intermediate stage of succession. Apparently its high effective leaf area allows White Pine to quickly and permanently over-top its early competition. Emergent Pines keep their foliage in an environment of light and moisture very like that of early succession while the trees around and beneath them compete through further stages of succession. If the Pines are numerous enough as pioneers, they can form a monospecific stand. Their accumulated litter is a poor seedbed and their deep shade deters the growth of seedlings and saplings in the understory. The stand of White Pine is highly resistant to invasion. It may not change in specific composition even over a length of time that is sufficient for normal successional change on a similar piece of land pioneered by some other species.

After discovering that White Pine and the climax of a successional series could form alternative communities that are stable over comparable spans of time, I searched local arboretums for trees with characteristics similar to White Pine's. In the Prospect Garden of Princeton University, I found a Giant Sequoia with 5.1 layers that let through only 0.7% of the incident light. Consequently I went to California to measure some Giant Sequoias in situ in Calaveras County Big Tree Park. Table 6.3 shows that Giant Sequoia indeed has far more layers than the other

TABLE 6.3. Layers of foliage of "saplings" less than 30 meters tall in a grove of Giant Sequoias, Calaveras County Big Tree Park, California.

Species	# Measured	% Light/ branch	% Light/ tree	# Layers ± SE
Giant Sequoia	19	48	0.7	7.4 ±0.5
White Fir	12	34	3.6	3.4 ±0.5
Incense-cedar	12	27	3.3	2.7 ±0.2
Pacific Dogwood	10	12	3.6	1.6 ±0.1

trees around and beneath it, and sheds deeper shade as well. Like White Pine, it also is a very efficient pioneer and is highly persistent at later stages of succession, even though it is less tolerant than the trees around it (Biswell, Buchanan, and Gibbens 1966). Incidentally, White Fir and Incense-cedar let through more light than the compensation point for most foliage. Like Gray Birch and Bigtooth Aspen, they are probably compromising light interception and water saving strategies.

The number of layers in trees should have predictive value even if the species that are measured are not identified. Table 6.4 shows measurements made at Finca Las

TABLE 6.4. Layers of foliage in a middle-altitude rain forest in Costa Rica. The stages of succession are relative rather than absolute. The height of the canopy is that of the tallest vegetation on the plot. Trees under 20 meters high were considered to be in the canopy if there was no vegetation anywhere above them. The mean number of layers is given ± its standard error, with the number of measurements in parentheses.

| | Height of canopy (m) | # Layers ± SE (# measured) in: | | |
Stage		Canopy	Understory (<20m)	Ground cover (<2m)
Old beanfield-pasture	2			5.2 ±0.3(50)
Early forest invasion	25	3.5 ±0.4(15)		2.7 ±0.2(19)
Second growth	40	4.1 ±0.4(37)	2.5 ±0.3(16)	1.3 ±0.1(22)
Late second growth	45	3.8 ±0.2(47)	1.6 ±0.2(18)	1.4 ±0.1(40)
Primary forest	50	3.5 ±0.7(7)	1.6 ±0.1(10)	1.1 ±0.1(20)

Cruces, near San Vito de Java, Puntarenas Province, Costa Rica. The plots are classified by successional stages which represent a consensus gleaned from legends associated with courses given by the Organization for Tropical Studies. I could not reconstruct accurate and reliable histories for any but the youngest plots, but the heights of the tallest trees in each plot agree with the guessed successional stages. The primary forest is impressive, but I doubt that it is pristine because there is some evidence that high grade lumber trees have been removed. In Table 6.4 there is an obvious gradient in the number of layers of plants from canopy through understory to ground cover. At all levels the number of layers tends to decrease with succession, with one exception. The lower number of layers in the canopy of the early forest invasion is due to trees like *Cecropia, Ochroma, Piper,* and *Croton,* all large-leaved small trees with very little secondary branching, and with soft or hollow stems. They look and behave like overgrown herbs, rather than like trees that are early in forest succession. In the wetter tropics, vegetational growth in early succession is so vigorous that no tree can expect to have more than its top layer of leaves above the general tangle. Therefore the advantages of the multilayer in the earliest stages of succession can only be realized in drier areas. The relatively high number of layers in the canopy of late-successional stages is probably due to the high death rate among trees that are destroyed by vines and epiphytes.

DISCUSSION

I have had difficulty persuading some people that trees that look as similar superficially as White Pine and Eastern Hemlock actually have different numbers of layers. Both appear to have many branches, but the distribution of foliage within branches is more even and sparse in White Pine. The needles of Eastern Hemlock are dense throughout the upper branches but only at the ends of all shaded branches. Thus the difference between projection of branch and tree is much greater for White Pine than for Hemlock (Figure 6.1).

The developmental ideas of Wiesner (*fide* Büsgen and Münch 1929) provide a mechanism by which trees could adjust their leaf arrangement to fit the optimal multilayer's strategy, including dense and uniform coverage in the lowest layer of a tree with few layers. He observed that trees self-prune branches that are dimly lit, and that branches that are only a little less dimly lit grow very little, resulting in their leaves being closer together and forming a more continuous cover. Conversely, if bare branches of some trees are well lit, they produce epicormic sprouts (Kramer and Kozlowski 1960). Wiesner's light measurements were made by timing exposures of photographic paper to reach a certain darkness. He assumed that the critical light levels were those that just allowed a branch a net gain in photosynthates. His work could profitably be repeated with a more accurate measurement of shading and corresponding measurements of the compensation point to test his physiological notions. If these ideas are physiologically realistic, they provide limited flexibility for the leaf arrangement of trees with a small number of layers.

It would clearly be advantageous for individual trees to have some developmental flexibility in their number of layers. A monolayer is adaptive for invasion of a shaded understory, but once the invader reaches the canopy it could fix more energy as a multilayer. Examples of such flexibility have already been given in Chapter 4 for oaks and firs. Nearly all broad-leaved seedlings are perforce monolayers; hence multilayered herbs have an advantage over seedlings of broad-leaved trees in the earliest stages of old-field succession. However most trees gain some semblance of the adult leaf distribution as saplings. Thus most species remain typically either monolayer or multilayer soon after they become independent of the food supply carried in their seeds.

A. Branch of White Pine.

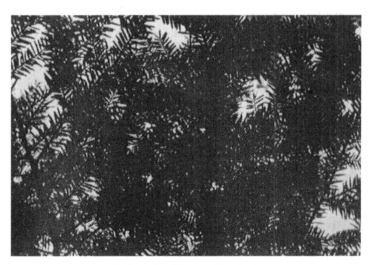

C. Branch of Eastern Hemlock.

FIGURE 6.1. Projections of branch and tree for White Pine and Eastern Hemlock.

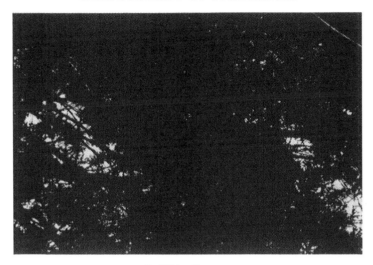

B. White Pine tree.

D. Eastern Hemlock tree.

FIGURE 6.1.

Several species that are characteristically monolayers when initially grown in the shade can also grow in the open. Table 6.5 shows the morphological parameters of Flower-

TABLE 6.5. Layers of foliage of monolayers grown in a forest and in the open.

Species	Forest-grown		Open-grown	
	# Measured	# Layers ± SE	# Measured	# Layers ± SE
Flowering Dogwood	13	1.4 ±0.1	12	1.5 ±0.2
Sugar Maple	8	1.9 ±0.1	13	1.9 ±0.1
American Beech	16	1.5 ±0.1	10	1.9 ±0.2
Eastern Hemlock	13	1.6 ±0.1	11	1.8 ±0.1

ing Dogwood, American Beech, Sugar Maple, and Eastern Hemlock when grown in shade and when grown in the open. The species all put out more leaves when grown in the open, and their effective leaf areas are comparable to those of open-grown multilayers. However, they do not seem able to add many more layers when grown in the open.

Three of the species that I know well show so much variability that it is difficult for me to measure a characteristic number of layers for any of them: Sassafras, Black-gum, and ash. It is comforting to note that these are also species to which foresters have had difficulty assigning tolerances.

Open-grown Sassafras has numerous, large, over-lapping leaves. Although the measurements show that Sassafras has several layers, my impression on looking at an open-grown tree is that the leaves are concentrated in a thin outer shell. Sassafras trees that have persisted until surrounded by forest have much sparser foliage. Sassafras remains enigmatic to me, although I suspect that the enigma might be resolved if I knew more about its bio-chemical relations with other species (Chapter 3).

Blackgum occurs in two distinct forms, one a multilayer with profuse horizontal layering, the other a monolayer with but one terminal layer. The monolayered form is particularly common among root sprouts and is found only in the understory. Most of the Blackgums in the vicinity of Princeton, New Jersey, are multilayers, but in Mammoth Cave National Park, Kentucky, monolayered Blackgums share the understory of an oak-hickory forest with Flowering Dogwoods. As befits their morphology, the Kentucky Blackgums are found beneath denser canopies than their New Jersey counterparts. The variations in tolerance of Blackgums are at least partly explained by their variations in morphology.

When grown in deep shade, ash saplings each year put out a pair of terminal leaves on a single stem. If these saplings are exposed to sunlight, they quickly put out lateral branches and several layers of leaves. Therefore a shaded ash may exist for years as a sparse monolayer ready to take rapid advantage of a sudden opening in the canopy, but unable to reach a closed canopy before the ratio of leaf area to supporting tissue becomes critically small. If tolerance is viewed as a species-specific charac-teristic, then the behavior of ash is puzzling. However if tolerance is related to morphology, we can understand why young ash saplings can be numerous in the densest forest without allowing ash to successfully invade the canopy (Büsgen and Münch 1929, Okali 1966).

The same argument applies to differences between species as well as to the developmental pattern within one species. For a monolayer the ratio of productive leaf area to nonproductive supporting tissue decreases rapidly as the tree gets larger. Thus there is a limit to the size of monolayered trees when the tree has grown until all of the energy fixed by the leaves is used to maintain the roots, trunk, and branches. Of course the same argument applies to a multilayer, but the ratio of leaf area to supporting tissue decreases more slowly since new branches allow new layers of leaves. Therefore the maximum size of mono-layers may be smaller than that of multilayers on the same site, leading to a forest with multilayers in the canopy and an understory of monolayered trees that characteristically never reach the canopy.

SUMMARY

Accurate tests of the predictions made in Chapters 4 and 5 demand a measure of exactly how many layers a given tree has. This measure is provided by the ratio of the logarithms of the amount of light that penetrates the whole tree and the amount that penetrates a single branch. That is how many like branches, independently distributed, would equal the whole tree.

With this measurement the number of layers is found to decrease from canopy to understory and from early- to late-successional stages in eastern deciduous forests of New Jersey, western coniferous forests of the Sierra of California, and middle elevation rain forests of Costa Rica. Xeric climax forests have more layers than mesic climax forests, and their trees apparently let through enough light to allow another layer of paying leaves, compromising light interception with water savings. The effective leaf area, the number of layers times the projection of leaves in each layer, decreases in the late stages of succession, suggesting a decline in gross productivity.

White Pine and Giant Sequoia are multilayered but shed deep shade. If such trees are numerous enough as pioneers, they can form a monospecific stand that is highly resistant to invasion over a length of time that is sufficient for normal successional change on a similar piece of land pioneered by some other species. Moreover if they are numerous and long-lived enough, they can quickly invade openings caused by the death of older trees and persist in a climax forest.

Within some single species there is limited flexibility in adjusting the number of layers to be adapted to a variety of environments, and there is corresponding variation in their tolerance in the different environments. However, monolayers can only approach and never equal the strategy of multilayers when grown in the open. Conversely, multilayers grown in the shade cannot adopt as efficient a morphology as monolayers can.

This limitation of flexibility and an additional energetic constraint lead to the prediction of macro-layering in some forests. Monolayers have less leaf area per unit of non-productive supporting tissue than multilayers of the same size. Therefore the maximum energetically possible size of monolayers is less than that of multilayers on the same site, resulting in a multilayered canopy with a permanent mono-layered understory.

CHAPTER SEVEN

Speculations on the Shapes
of Tree Crowns

The shape of a tree in the forest is largely determined by the shape of the space that it fills. However, many species attain characteristic shapes when grown in the open. So there is an inherited, developmental tendency for each tree to attain a particular shape, which may be modified by the environment where the tree grows. The shape of a tree determines the total amount of light that it intercepts and limits the strategies of leaf placement. From the combined constraints of development and light interception, I shall generate a table that suggests optimal shapes and growth patterns for monolayered and multilayered trees.

First it is necessary to review some constraints that are imposed by development or by the environment but are unrelated to light interception. Then, although there is an infinity of possible shapes, the various shapes of trees can be distinguished by three parameters: absolute size, ratio of height to width, and convexity. The ratio of height to width is constrained by the cost of horizontal branches, which is greater than that of vertical branches of the same length. This ratio, and the convexity of a tree, are adjusted to give an optimal combination of growth and light interception in a given environment.

Because the ideas in this chapter have not yet been tested, there is more formalism than is needed to derive a table of optimal shapes and growth patterns for mono-layered and multilayered trees. The sections *Parameters of the Shapes of Trees* and *Interception of Light by the Whole Tree* are expanded beyond the immediate needs of the chapter because they suggest measurements that might be used to test the theory.

CONSTRAINTS ON THE SHAPES
OF TREES

Some patterns of development invariably determine the shape of a tree. If there is no branching from the central trunk and if leaves are borne in a tight terminal spiral or rosette, then the only possible shapes are parasols and lollypops. Palms, most other monocotyledonous trees, and tree-ferns all show some variant of this developmental pattern. Therefore parasols and lollypops will often be found in environments that favor trees that have scattered vascular tissue which might insulate sensitive tissues against severe solar radiation and surface heating. An excurrent branching pattern, with side branches going straight out from the main branch, combined with a tendency for terminal buds to suppress the growth of lateral buds, produces a conical shape. Most gymnosperms combine excurrent branching with terminal dominance. Hence conical shapes will be found where the environment favors needle-leaved evergreens.

Other developmental patterns are more flexible. For example, deliquescent branching can produce virtually any regular shape by varying the angle of branching and the degree of terminal dominance (Büsgen and Münch 1929).

In some cases the shape of a tree has an obvious adaptive significance that has no relation to the light intercepted. The pagoda-like cones of northern conifers shed much snow and spread the weight of what remains evenly among the branches. However, the conical shape alone is not enough to adapt trees to snow. Aspens and birches are conical but their erect branches when splayed by the weight of snow accumulate even more snow. The remarkably flat branches of trees on tropical savannahs, alpine cloud forest emergents, timberlines caused by Chinook winds, and large trees chronically exposed to salt spray may find a common significance in resistance to drying winds. (In the case of salt spray the drying effect is osmotic.) Most leaves in such branches are in the wake of other leaves. The shape of the whole branch may reduce turbulence and produce a more laminar flow of air past the leaves. These speculations require documentation, but there is clearly an adaptive analog for the flattened branches that are caused directly by chronic exposure to wind.

The preceding factors must be kept in mind when discussing patterns of tree shape, because they will certainly interact with any effect of light interception, and they may even overwhelm it.

PARAMETERS OF THE SHAPES OF TREES

Trees can take so many shapes that it would take an infinite number of parameters to express all of them accurately. However there is a simple equation that can represent a wide variety of shapes in transverse section, with only a few easily interpreted parameters. That equation is

$$x^a + (by)^a = c^a, \tag{7.1}$$

where x and y are variable Cartesian coordinates and a, b, and c are constants. This equation behaves nicely in the quadrant where x and y are positive. When $a = 1$, it is a straight line; when $a = 2$, it becomes convex and bows out from the origin to form part of an ellipse; if a is larger than 2, it is more convex, becoming rectangular as a approaches infinity. The ratio of height to width is measured by b, and the absolute size by c.

The transverse section of a tree can be generated by reflecting this equation into the other three quadrants. If different values of a and b are allowed for quadrants above and below the x axis, then the equation $x^a + (by)^a = c^a$ can represent any tree's shape from a cylinder, ellipsoid, or tapered spindle, to a mushroom, lollypop, or inverted ice-cream cone.

We can therefore represent the shape of a tree with only three parameters, each having a simple geometrical interpretation, size, ratio of height to width, and convexity. If these parameters were independently related to corresponding parameters of development like growth, terminal dominance, and branching angle, then optimizing tree shapes would be a simple problem in calculus. However, a tree's convexity is probably affected by its ratio of height to width, which surely changes in many species as the tree grows. Thus developmental changes in the parameters of tree shape are as important as the absolute values of the parameters.

THE COST OF WOOD IN VERTICAL
AND HORIZONTAL BRANCHES

The woods of different trees vary greatly in caloric content and strength. Hardwood has heavily lignified cell walls, or extensive deposits of pitch and resin in the compression wood of conifers (Jane 1956). Hardwood is heavy even when dried; it is very strong and rigid, but somewhat brittle. Softwood has a much lower caloric content per unit of volume; it is much lighter than hardwood when dried, but when it is green there is less difference in weight (Section 14 in Forbes 1955). Softwood is flexible and elastic, but of course its thin-walled cells are easily ruptured when

107

a branch is flexed beyond its elastic limit (Jane 1956). A given volume of hardwood has a higher caloric cost to the tree than softwood. In return for its higher cost, hardwood is better able to support the tree and its leaves against the constant compression and torque of gravity. Variable winds call for compromises between the rigid strength of hardwood and the elastic flexibility of softwood, but these compromising strategies have not been adequately studied.

Horizontal branches support their weight against a gravitational torque much greater than that on the trunk or on branches that are nearly vertical. Therefore a horizontal branch has a higher caloric cost than a vertical branch of the same length. The cost must be paid in either a greater cross-sectional area or stronger wood, both of which increase the weight of the branch. The hollow, air-filled, trussed structure seems little used by trees with secondary growth, although examples are found among the monocotyledonous trees. Furthermore, the gravitational torque that a horizontal limb must resist increases as its length times its weight; the strength of a vertical branch need only increase as its weight, as long as the branch is not so tall that it is flexed by its own weight (Greenhill 1881). Therefore, as a tree grows, the relative cost of lengthening horizontal branches becomes increasingly greater than that of lengthening vertical branches. Thus for mechanical reasons, the most economical ratio of height to width increases as a tree grows.

INTERCEPTION OF LIGHT BY
THE WHOLE TREE

Jahnke and Lawrence (1965) have examined the relation between the shape of a tree and the total amount of light that it intercepts during the day. Since the only figure that their analysis allowed was a cone of varying height, it is not surprising that they showed that the higher the cone, the more light it intercepted during the course of the day. The following is a more generalized account than that of Jahnke and Lawrence. An infinity of possible shapes is allowed, and the average projection over the day is calculated.

We start by slicing the tree into very thin transverse sections parallel to the daily path of the sun. Then we calculate a one-dimensional projection for each two-dimensional slice. Finally, by adding the projections of all the slices, we get the two-dimensional flux of light intercepted by a three-dimensional tree. Each slice of the tree is a generalized convex figure, which can be described in polar coordinates as follows (see Figure 7.1). An arbitrary point inside the figure is chosen as the origin of polar coordinates, and a straight line is drawn at an arbitrary reference angle θ. A tangent to the figure is erected perpendicular to this line. The distance from the origin to the tangent line is called

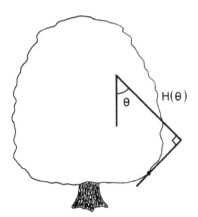

FIGURE 7.1. Description of the shape of a tree.
$H(\theta)$ is the distance from the origin of polar coordinates to a tangent on the edge of a slice through the middle of the tree. See text for further discussion.

109

$H(\theta)$. Although a given function $H(\theta)$ does not uniquely determine a particular figure, each figure has a unique $H(\theta)$.

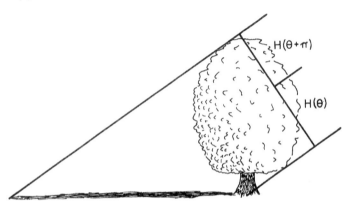

FIGURE 7.2. Projection of a tree.
The projection is the amount of light flux intercepted, not the length of the shadow on the ground. Thus the projection $=$ $H(\theta) + H(\theta + \pi)$ for a thin slice from the middle of the tree.

The projection of the tree is $H(\theta) + H(\theta + \pi)$ for a light shining from an infinite distance at an angle of $\theta + \pi/2$ or $\theta + 3\pi/2$ (see Figure 7.2). The average projection of the figure as the light moves around in a half circle is then:

$$\int_0^\pi [H(\theta) + H(\theta + \pi)]d\theta = \int_0^\pi H(\theta)d\theta + \int_0^\pi H(\theta + \pi)d\theta$$

$$= \int_0^\pi H(\theta)d\theta + \int_\pi^{2\pi} H(\theta)d\theta$$

$$= \int_0^{2\pi} H(\theta)d\theta.$$

This integral can be directly related to the geometry of the figure by using a proof found in Kendall and Moran (1963). Moving the point of tangency of the tangent line a small distance dP along the perimeter of the figure will

110

move the line $H(\theta)$ through an angle $d\theta$. Since θ is measured in radians, then $dP = H(\theta)d\theta$ (see Figure 7.3). Tracing the entire perimeter of the figure with the point of tangency moves $H(\theta)$ in a full circle. Mathematically:

$$\int_0^{2\pi} H(\theta)d\theta = \int_0^{2\pi} dP = \text{perimeter of the figure.}$$

If the transverse section of a tree is not convex, then this proof holds for the smallest convex figure that fully encloses the tree.

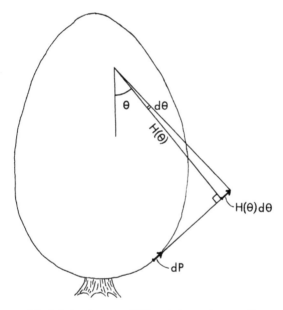

FIGURE 7.3. Relation between $H(\theta)$ and the perimeter of a convex tree.

For those who worry about such things, there is a mathematical subtlety here. I move the tangent point a distance dP to define $d\theta$ by $dP = H(\theta)d\theta$; I do not move $H(\theta)$ through an angle $d\theta$ to drag the tangent through a distance dP. To get the perimeter I really integrate piecewise around the perimeter rather than piecewise over θ, since the perimeter defines $H(\theta)$ rather than vice versa. This procedure is legal for my admissible curves, namely closed, smooth, convex, single-valued curves.

111

The projection of a whole tree is then the sum of the convex perimeters of slices in the plane of the sun's path, each multiplied by the thickness of the slice. Another geometrical interpretation is as follows. Enclose the crown of the tree in a stretched bag. Find the area of each and every flat section of the bag, and multiply it by the sine of the angle between that section and the plane of the sun's path. Add the results for all parts of the bag and you have a weighted convex surface area, which is equal to the area obtained by summing convex perimeters of slices in the plane of the sun's path, each multiplied by the thickness of the slice. In other words, the projection of a tree over the day is proportional to the convex surface area of its crown, counting fully any parts of the surface that the sun strikes at right angles, and discounting parts that the sun always strikes at lesser angles. Jahnke and Lawrence's (1965) measurements demonstrate a special case of this result, but for a tree that covers a fixed amount of ground, a rectangular transverse section, a cylinder, intercepts more light over the day than a triangular section, a cone, of the same height.

Few trees are actually rectangular in transverse section; there are other constraints on their shape beside light interception. For example, we might want to minimize the distance that raw materials and products of photosynthesis are transported, to minimize the bending moments exerted on branches by gravity and wind, to add developmental constraints, and then to maximize the total interception of light that varies in intensity as the sun strikes from different angles. This problem is similar to classical isoperimetric problems of the calculus of variations (Courant and Hilbert 1953), in which one looks for the function that maximizes the value of a given integral while keeping the value of another integral constant. I have set the problem in this form, but I have been unable to give Euler's equa-

112

tion in explicit form or to guess a proper solution. For the present I adopt the less elegant strategy of generalizing the results of this section as qualitative axioms.

In order to intercept the maximum flux of light over the day, the perimeter of a vertical section of the tree must be large, particularly those portions of the perimeter that are perpendicular to the directions from which the most intense light comes during the day. Local concavities in the shape of the tree add nothing to its light interception. Intuitively, this means that a big, fat tree intercepts more light than a little, scraggy one. However, the proof gives a quantitative and measurable meaning to fatness and scragginess.

PATTERNS OF OPTIMAL SHAPE IN TREES

The patterns of tree shape that are optimal for successional strategies are listed in Table 7.1.

TABLE 7.1. Adaptive patterns in the shapes of tree crowns.

Property	Multilayer	Monolayer	Persistent multilayer
Shape as a sapling	tall and thin	flat and spreading	tall and thin
Wood	soft	hard	soft sapwood hard heartwood
Change in ratio of height to width	fast increase	slow increase	decrease
Shape of large trees growing in the open	cone	tall ellipsoid to cylinder	flat ellipsoid to cylinder

Early-successional trees should be multilayered. Since early succession is in some sense a race to form a canopy, fast-growing softwoods are favored over stronger hardwoods. Growth in height is favored over growth in width;

113

hence trees should initially be tall and thin. As the tree gets larger, lateral branches become increasingly more expensive than vertical branches. Since softwood is relatively weak, lateral growth becomes extravagant, and the ratio of height to width should increase rapidly as the tree grows. The tall outline also allows layers of leaves to be far enough apart so that one layer does not completely eclipse the sun from leaves in the next layer and the tree can take full advantage of the multilayered strategy. The high productivity of a multilayered tree implies that large amounts of nutrients and photosynthetic products must be transported. The distance that they are transported is less if the convexity is small. Softwood branches that stick out of the outline of the tree would also be sensitive to storm damage. Therefore the early successional multilayer should be minimally convex, but still convex so that the whole perimeter adds to light interception. The minimally convex figure that is consistent with most developmental patterns is a cone.

Trees that invade the understory late in succession should be monolayered. In a uniform stand, most of the light in the understory comes from near the zenith because the path of light through the canopy is shorter when the sun is at the zenith than when the sun strikes the canopy obliquely. Therefore the foliage of understory trees should be flat and spreading, a shape that also befits a monolayered distribution of leaves. To support extensive lateral branches requires the rigid strength of hardwood, even in a sapling. The relative cost of lateral branches increases with growth, but at a lesser rate for hardwoods than for weaker softwoods. Thus the ratio of height to width should increase slowly as the tree grows. The hardwood is less sensitive to storm damage, and the monolayered foliage makes food at a lower rate. Consequently economy of supporting and conductive tissue is not as important for monolayered trees as it is for multilayered trees. The monolayer can then be convex to intercept as much light as possible during the day.

Some multilayered trees persist in a climax forest that is actually a patchwork of successional stages. Such forests are found where regular, spotty deaths are caused by drought, seasonal flooding, or shading by vines. The persistent multilayer invades small openings in the forest and should have a mixed strategy. When young it must race to the canopy with early-successional multilayers. It therefore should be tall, thin, and multilayered, with softwood. After reaching the canopy it should spread out rapidly to dominate the opening. The ratio of height to width decreases with age. Extensive lateral growth in the canopy demands the supporting strength of hardwood. Thus its new growth should be pithy, its live and growing sapwood should be softwood, but the dead and supporting heartwood should be hardwood.

Figure 7.4 summarizes these predictions. Early-successional trees should be tall, thin, and conical, and should have softwood and a multilayered leaf distribution. Monolayers should be flat and spreading as saplings, tall ellipsoids filling out to cylinders as they reach maturity; they should have hardwood. Multilayers that persist in the climax should be tall cones as saplings, flattening and spreading as they grow to become mushroom-shaped or perhaps cylindrical at maturity; they should have pithy twigs and soft sapwood but hard heartwood.

SUMMARY

Trees show great variation and flexibility in their shapes. The adaptive significance of some shapes can be guessed easily, for example the flat branches of trees exposed to drying winds or the snow-shedding pagodas of northern

115

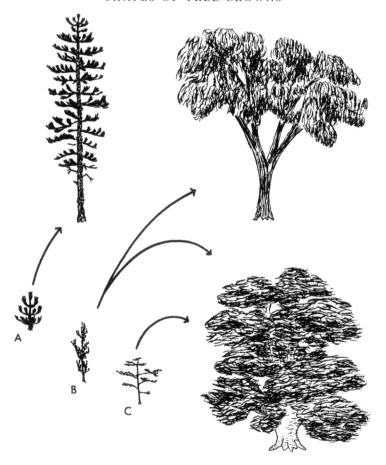

FIGURE 7.4. Patterns in the shapes of trees.
A. Early-successional multilayer.
B. Persistent multilayer.
C. Monolayer.

116

conifers. The shape of a tree is also part of its strategy of light interception or successional status.

The shape of a tree is roughly defined by three parameters: its absolute size, its ratio of height to width, and its convexity. The ratio of height to width should be adjusted to provide a balance between growth in height to reach the canopy, and growth in width to intercept light and occupy sufficient space in either the understory or the canopy. The appropriate balance depends on the environment in which the tree is growing as well as the strategies adopted by its neighbors, and it is subject to the constraint that as the tree grows, horizontal branches become increasingly more expensive per unit of length than vertical branches. The total light that a tree intercepts during the day is a function of the convex perimeter of its vertical section. In particular, a cylinder has the largest projection of any shape that covers a fixed area on the ground, and local concavities in the shape of a tree add nothing to the total light intercepted. However the greater the convexity of a tree the greater must be the amount of supporting and conductive tissue that is devoted to a unit of its productive perimeter.

From these considerations, the following predictions are made. Early-successional trees should be tall, thin, and conical, and should have softwood and a multilayered leaf distribution. Monolayers should be flat and spreading as saplings, tall ellipsoids filling out to cylinders as they reach maturity; they should have hardwood. Multilayers that persist in the climax should be tall cones as saplings, flattening and spreading as they grow to become mushroom-shaped or perhaps cylindrical at maturity; they should have pithy twigs and soft sapwood but hard heartwood.

CHAPTER EIGHT

On the Relation between Theory and Reality

The ultimate test of a theory in empirical science is whether it works. The previous chapters have resolved some venerable questions, at least to my satisfaction. Although there is a voluminous literature on the exquisite adaptations of plants to extreme environments, I have found that many botanists are uncomfortable with discussions of adaptive strategies on a broader scale, especially discussions based on a single factor as ubiquitous as light. Therefore I offer précis of the adaptive argument and of the uses of an incomplete theory, pointing out appropriate examples from the preceding chapters. In particular, the theory gives some new insights into successional patterns of productivity, stability, and diversity.

THE ORIGIN AND MAINTENANCE OF ADAPTIVE PATTERNS

Adaptive differences between species develop during evolution by natural selection. If a large number of differently adapted plants are able to invade a variety of environments, then species that have an adaptive advantage over others in a given environment also have a competitive advantage. Therefore the composition of a community tends toward the association of available species who have the highest efficiency of occupying the several patches of the environment. Thus a contemporaneous adaptive correlation arises between environment and phenotype, analogous to the correlation between environment and adaptation during evolution.

The adaptive argument starts by examining the effect of each phenotype on an appropriate measure of success in a given environment: reproductive output, growth, or persistence of the dominant phenotype in the face of potential invasion by another phenotype. This part of the argument is the same whether one is talking about the evolution of diverse adaptations or the contemporaneous maintenance of an adaptive pattern. I have been told that Eastern Redcedar and Gray Birch are conical multilayers because their developmental patterns are rigid, not because any particular form or leaf distribution is adaptive to early succession. My response rephrases the argument to ask why a tree with a particular developmental pattern is found where it is. Some species, like Blackgum (Chapters 3 and 6), are flexible enough to have many layers when early-successional but fewer later. Others are characteristically multilayered and early-successional, like Gray Birch, or monolayered and late-successional, like American Beech (Chapter 6). Sometimes a tree adopts the appropriate strategy from a list of possible strategies; indeed evolutionary differences between species can only arise in this way. But adaptive patterns can be maintained even among inflexible species if nature picks those trees that have strategies appropriate to each environment. Flexibility and specialization are adaptive in their own right (Levins 1968), and the significance of flexible leaf distribution is an intriguing and untouched problem.

I have also been told that the early-successional status of trees such as Gray Birch has nothing to do with their shape or leaf geometry, but is dependent only on the wind-dispersal of their seeds; or that another tree's root structure allows late-successional status. Some strategies obviously preclude others. For example, the large seedling

and root hypocotyl that are needed to penetrate accumulated leaf litter (Coile 1940) require a heavily provisioned
seed, which cannot be dispersed by wind. The shapes of
trees are adaptive compromises of conflicting strategies
(Chapter 7). However, as long as a particular means of
dispersal, or an efficient root system, or a certain shape
does not preclude a given leaf geometry, the characteristics
should evolve together, and trees that typify a particular
stage in succession should show a constellation of adaptations to that stage.

An adaptive argument is inherently deterministic, and
its predictions may fail or be ambiguous for historical
reasons. The locally available species may not encompass
a large enough range of variation. For example, when a
large area of deciduous forest in the northeastern United
States is cleared for lumber, the early second growth
is often those trees of the original forest that sprout
quickly from stumps, rather than species that are the
pioneers of succession in old fields. Again there may be a
conflict between an efficient association and a less efficient
association of individually more efficient species. An example is provided by the alternatives of a full successional
series or a succession dominated from the start by White
Pine (Chapter 6).

A testable adaptive pattern describes a correlation between phenotypes and characteristics of the environment.
Thus measurements can either refute or provisionally confirm an adaptive theory even in a flora that is not well
known taxonomically. For example, the relation between
layering and succession applies to tropical rain forests
even though the names of the species were ignored when
the data were gathered (Chapter 6).

USES OF AN INCOMPLETE THEORY

What are the uses of a theory of forest structure that is based primarily on light interception? It is obviously incomplete because there are always other factors operating.

To the extent to which a single factor either overrides other factors or is independent of them, the theory can make accurate predictions. A frontal assault on the first factor in a multidimensional problem may show that many of the presently known patterns can be understood in terms of that factor alone. Many of the generalizations of forest succession can be explained by the number of layers of leaves in trees at each stage (Chapter 4). Such a posteriori predictions may look like teleology or mathematical obfuscation. But even if the predicted patterns have long been known, the theory replaces nebulous intuition with an explicit list of testable assumptions, and it can also indicate hidden assumptions. For example we cannot accept uncritically the assumption that stable age distributions must replace senile ones (Chapter 5).

Departures from predictions based on a single factor may point conspicuously to the next factor to be incorporated into the theory. The biochemical warfare of Sassafras leaves was discovered because Sassafras is a conspicuous exception to a prediction based on light alone (Chapter 3). Further departures from predictions may remain as minor embarrassments to the theory or as grounds for interesting speculation. Perhaps plants like Sassafras have evolved a mechanism that allows them to let through enough light to support their offspring while preventing invasion by species that are better adapted to shade. Another such speculation is the interpretation of large-leaved, early-successional monolayers of the wet tropics as overgrown herbs, adapted to avoiding shade rather than to enduring it (Chapter 6).

A theory based on a single factor can identify parameters that are simple but sufficient measures of the effect of that factor. An obvious example of such a parameter is the number of layers in a tree. One striking characteristic of the number of layers in a tree is its low variance. I can sometimes distinguish differences between species with as few as 10 measurements of the number of layers in each, whereas I would need hundreds of measurements of each of the same species to show comparable differences in canopy preferences (Chapters 6 and 3). Perhaps if ecologists could identify and correlate a few more biological parameters with such low variance, we might be able to propose laws as exact as those of primitive physics.

Finally first-order theories set some limitations on second-order theorizing, shed new light on old problems, and may even predict totally new patterns. Most of the current theories of succession have been based on plausible analogy rather than on rigorous analysis. Once succession is viewed as a progressive developmental process, it is easy to assume that admirable qualities such as productivity, stability, and diversity increase as succession proceeds. Odum (1969) has reviewed both data and theory of succession and finds that none of these qualities shows a monotonic increase during succession. Hence it behooves me to ask what a rigorous, analytic, though incomplete theory of forest succession has to say about productivity, stability, and diversity.

PRODUCTIVITY

Odum (1969) finds that actual patterns of productivity in succession are consistent with the simple but rigorous theory proposed by Lindeman (1942). As long as the production of a community outweighs its respiration, new organic matter will accumulate. If the production of this new matter outweighs its respiration, net productivity will increase. Eventually the new matter must cost more than it produces, either because it blocks light from part of the community or because it is devoted to tasks other than photosynthesis. Net productivity must then decline until finally respiration equals production in a steady state with no net increase or decrease of organic matter, no net production. The theory of forest structure is entirely consistent with this view. In addition it predicts that even gross productivity should decrease if monolayered trees reach the canopy late in succession (Chapter 5). It predicts that community productivity should depend less on the total flux of light intercepted than on the geometrical distribution of leaves that intercept it. Limited empirical support for this prediction is given in Figure 8.1. The average width of an annual ring is correlated with the number of layers in each of several species of tree, but there is no correlation, or perhaps even a negative correlation, with the amount of light that the tree intercepts. Whittaker (1966) finds no correlation between accurate measures of productivity and total light interception. The relation between layering, or perhaps effective leaf area (Chapter 6), and productivity is a promising and practical area for future studies.

A corresponding theory could be developed for any community that uses a unidirectional resource. For example the most productive community would be an infinitude of layers of sparse and microscopic leaves with no nonproductive supporting tissue. Phytoplankton in turbulent water form such a community. Indeed some planktonic communities exceed the greatest productivity among terrestrial communities, partly because of their geometry and partly because of their rapid recycling of essential nutrients (Odum 1959).

123

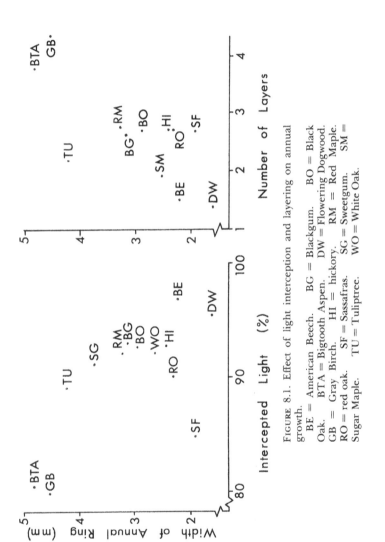

FIGURE 8.1. Effect of light interception and layering on annual growth.

BE = American Beech.   BG = Blackgum.   BO = Black Oak.   BTA = Bigtooth Aspen.   DW = Flowering Dogwood.   GB = Gray Birch.   HI = hickory.   RM = Red Maple.   RO = red oak.   SF = Sassafras.   SG = Sweetgum.   SM = Sugar Maple.   TU = Tuliptree.   WO = White Oak.

STABILITY

If stability is defined as persistence in time, then the statement, "Stability increases as succession proceeds," simply rephrases the definition of stability. To avoid this circularity, stability may be defined by a stationary age distribution in which reproduction and growth just balance mortality in each age class. An alternative and equally useful definition of stability is "resistance to invasion." Both of these definitions are reasonable, but the stabilities so defined may be mutually exclusive. It is impossible to design a multilayered leaf distribution that both allows regeneration in its own shade and prevents invasion by a more shade-adapted morphology (Chapter 6). The most efficient monolayer sheds complete shade, effectively preventing invasion, but also preventing its own continuous reproduction. Thus it precludes a stable age distribution. Thus if climatic and edaphic conditions allow a monolayered climax, it is resistant to invasion but its age structure is not locally stable, and need not be stable even globally. The same is true if a drier climate or drier soil allows a multilayered climax forest with a monolayered understory. In this case the understory is protected from a high heat load by the multilayered canopy above. The monolayer cannot reach the canopy because of its low ratio of leaf area to supporting tissue (Chapter 6), and it shades the forest floor so completely that continuous regeneration is impossible. However, in multilayered forests where isolated trees die frequently because of storms, predation, or parasitic vines and epiphytes, the climax is actually a patchwork of various successional stages. Clearings are continually opened to invasion, and the age distribution of trees may reflect a stable age distribution of openings in the forest. We can say therefore that stability increases as suc-

125

cession proceeds only if for different successions we use opposing definitions of stability.

DIVERSITY

The theory shows that there is only one way to monopolize light. Thus if monolayered trees reach the canopy in a climax forest, the diversity of trees in the forest must decrease through the later stages of succession. Then why are there so many species in some forests? The answer must involve other adaptations besides monopolizing light. First we should find some species that are relicts of an earlier successional stage. There might also be some species that depend on frequent deaths among trees of the forest to create a constant supply of new clearings to be invaded. Several species might reproduce themselves uniformly over the whole area copiously enough to have relatively stable age distributions. Finally there might be species that are characteristic of the climatic climax in an undisturbed area, relentless invaders that are pushed back occasionally by widespread, long-term secular changes in the environment. Within each of these categories we could have understory and canopy trees because of the energetic limitations discussed in Chapter 6.

To construct something like a periodic table of the species, only two types of data are needed, the age distribution of trees and their canopy preferences (Chapter 3). These data are given in Tables 8.1 and 8.2 for a forest on sandstone-capped uplands in Mammoth Cave National Park, Kentucky. Table 8.1 shows that Red Maple and ash are relicts of an earlier stage of succession, American Beech and Eastern Hophornbeam are invading, and except for Sugar Maple and Eastern Redcedar the rest appear to have relatively stable age distributions. Table

126

TABLE 8.1. Relative age structures of forest at Mammoth Cave National Park, Kentucky. The basal area of each species is given as parts per million of ground area. An area of 3263 m² was surveyed.

Species	Basal area (ppm)	Midpoint of each diameter class (cm)						
		0	5	15	25	35	45	50
White Oak	913	67	6	4	3		3	8
red oak	697	143		3	4	1	3	5
hickory	478	186	49	18		4	1	2
Tuliptree	449	13	9	6	1	1	2	3
Blackgum	323	224	195	19	4	1	1	
Red Maple	122	27	140	9		1		
Flowering Dogwood	92	613	152	2				
American Beech	87	6	11	2	3	1		
Sugar Maple	42		29	3				
Sassafras	16	80	9	1				
ash	15	37	34	1				
Eastern Redbud	10	78	7	2				
Eastern Hophornbeam	8	111	11	1				
Eastern Redcedar	1	20	1					

TABLE 8.2. Canopy preferences in an oak-hickory forest at Mammoth Cave National Park, Kentucky. Canopy preference (Chapter 3) is the log-transformed mean percent of skylight penetrating the canopy above a given species of sapling. It is followed by the larger deviation found by adding or subtracting the standard error of the mean of the log-transformed data.

Species	Canopy preference (% skylight)		Number of measurements
Sassafras	18	±3	36
Tuliptree	13	±3	24
hickory	8.1	±2.0	40
White Oak	6.3	±1.1	20
Red Maple	5.9	±0.7	86
red oak	5.6	±1.2	27
ash	5.4	±1.1	43
Eastern Hophornbeam	5.4	±1.2	45
American Beech	4.7	±1.4	19
Sugar Maple	3.8	±0.8	29
Flowering Dogwood	3.5	±0.5	134
Blackgum	3.3	±0.5	73

8.2 shows that hickory, Tuliptree, and Sassafras saplings are found only under relatively open canopies. Sassafras, Flowering Dogwood, Eastern Redbud, and Hophornbeam never reach the canopy, and the younger Blackgums are not growing out of the understory either. These results are summarized in Table 8.3. Of course Table 8.3 does not

TABLE 8.3. Strategies of tree species in Mammoth Cave National Park, Kentucky.

Strategy	Canopy	Understory
Successional species	Red Maple ash	Blackgum
Invaders of clearings	Tuliptree hickory	Sassafras
Stable species	White Oak red oak	Flowering Dogwood Eastern Redbud
Invaders in shade	American Beech	Eastern Hophornbeam

explain the diversity of trees in this forest, but it does reduce the formidable task of explaining the coexistence of fourteen species to the less formidable, though yet unsolved, one of explaining the coexistence of several pairs of species.

Competition between trees in its most general sense is competition to fill space, quickly in early succession but completely in late succession. In MacArthur and Wilson's (1967) terms, r-selection in early succession gives way to K-selection in the climax forest. Hence the species that occupies space most completely, to the exclusion of other species, should prevail in the climax forest. However, a monolayered climax forest cannot regenerate continuously, and multilayered climax forests must regenerate in a patchy fashion. In either case reproductive adaptations are extremely important for persistence. Reproductive strategies allow much more sympatric and

contemporaneous diversity than strategies of light interception, especially if the interactions between plants and
animals are considered (Janzen 1970). Any spatial diversity
in the environment, such as heterogeneity of soil or of
micro-climate, allows different species to monopolize the
available space in different places, and the resulting mosaic
of plants creates another level of spatial diversity to which
new plants may become exclusively adapted (Whittaker
1969). Further diversity may be generated, even in a uniform soil and climate, by predation, parasitism, and bio-
chemical competition and warfare between trees (Langford and Buell 1969). However to allow the coexistence of
several species, these agents must prevent monopolies and
"thin out" each species as predation does in Janzen's (1970)
model. Harper (1969) has already shown that such thin-
ning out is not a universal effect of predation. Theories on
these subjects are scarce, but critical to discovering how
tree species divide up their environment.

In a recent conversation, James Porter suggested that
sessile marine zoophytes, especially corals that depend on
photosynthesis by their enclosed algae, should conform to
the theories of Chapters 5 and 7. Indeed, corals are often
rounded or branched in the bright light of shallow waters
but flattened in the darkness of canyons, depths, and caves.
(P. J. Roos, 1967, Growth and occurrence of the reef coral
Porites astereoides Lamarck in relation to submarine radiance
distribution, Doctoral dissertation, Universiteit van Am-
sterdam, Drukkerij Elinkwijk, Utrecht; T. F. Goreau, 1959,
The ecology of Jamaican coral reefs. I. Species composition
and zonation, *Ecol.* 40: 67–90.)

Nomenclature

For the convenience of laymen and zoologists, I have used vulgar names throughout the book. My usage follows C. F. Brockman, 1968, *Trees of North America: A Golden Field Guide,* Golden Press, New York, except for Tuliptree and Blackgum, which I cannot bear to call Yellow Poplar and Black Tupelo. The names are capitalized when I refer to a particular species. The names of all plants that are only given vulgar names in the text are listed below.

American Beech	*Fagus grandifolia*
ash	*Fraxinus*
aspen	*Populus tremuloides, P. grandidentata*
Bigtooth Aspen	*Populus grandidentata*
birch	*Betula*
Blackgum	*Nyssa sylvatica*
Black Oak	*Quercus velutina*
Black Walnut	*Juglans nigra*
Butternut	*Juglans cinerea*
cedar	*Cupressaceae*
conifer	Order *Coniferae*
cottonwood	*Populus deltoides, P. trichocarpa, P. fremonti, P. sargenti, P. angustifolia*
Eastern Hemlock	*Tsuga canadensis*
Eastern Hophornbeam	*Ostrya virginia*
Eastern Redbud	*Cercis canadensis*
Eastern Redcedar	*Juniperus virginiana*
elm	*Ulmus*
European Beech	*Fagus sylvatica*
fir	*Abies*

131

Flowering Dogwood	*Cornus florida*
Giant Sequoia	*Sequoia gigantea*
Gray Birch	*Betula populifolia*
gum	*Nyssa, Liquidambar*
gymnosperms	Class *Gymnospermae*
hemlock	*Tsuga*
hickory	*Carya*
Incense-cedar	*Libocedrus decurrens*
juniper	*Juniperus*
larch	*Larix*
Lettuce	*Lactuca sativa*
Loblolly Pine	*Pinus taeda*
maple	*Acer*
Maple-leaf Viburnum	*Viburnum acerifolium*
Mayapple	*Podophyllum peltatum*
Northern Red Oak	*Quercus rubra*
oak	*Quercus*
Pacific Dogwood	*Cornus nuttallii*
palm	*Palmae*
pine	*Pinus*
poplar	*Populus*
Red Maple	*Acer rubrum*
red oak	Section *Erythrobalanus* of *Quercus*
Redwood	*Sequoia sempervirens*
Sassafras	*Sassafras albidum*
Shagbark Hickory	*Carya ovata*
Southern Red Oak	*Quercus falcata*
spruce	*Picea*
Sugar Maple	*Acer saccharum*
Sweetgum	*Liquidambar styraciflua*
tree-fern	*Cyatheaceae* and *Dicksoniaceae* of Class *Filicinae*
Tuliptree	*Liriodendron tulipifera*

walnut	*Juglans*
White Fir	*Abies concolor*
White Oak	*Quercus alba*
White Pine	*Pinus strobus*
willow	*Salix*

Bibliography

Anderson, M. C. 1964a. Studies of the woodland light climate, I. The photographic computation of light conditions. *J. Ecol.* 52: 27–41.

――――. 1964b. Light relations of terrestrial plant communities and their measurement. *Biol. Rev.* 39: 425–486.

――――. 1966. Stand structure and light penetration, II. A theoretical analysis. *J. Appl. Ecol.* 3: 41–54.

Baker, F. S. 1950. *Principles of Silviculture.* McGraw-Hill, New York.

Bard, G. E. 1952. Secondary succession on the piedmont of New Jersey. *Ecol. Monog.* 22: 195–216.

Baumgartner, A. 1967. The balance of radiation in the forest and its biological function. Pp. 743–754 in S. W. Trompe and W. H. Weihe, eds., *Biometeorology 2.* Pergamon Press, New York.

Biswell, H. H., H. Buchanan, and R. P. Gibbens. 1966. Ecology of the vegetation of a second-growth *Sequoia* forest. *Ecol.* 47: 630–634.

Bormann, F. H. 1956. Percentage light readings, their intensity-duration aspects, and their significance in estimating photo-synthesis. *Ecol.* 37: 473–476.

――――. 1958. The relationships of ontogenetic development and environmental modification to photosynthesis in *Pinus taeda* seedlings. Pp. 197–215 in K. V. Thimann ed., *The Physiology of Forest Trees.* Ronald, New York.

Buell, M. F., A. N. Langford, D. W. Davidson, and L. H. Ohmann. 1966. The upland forest continuum in northern New Jersey. *Ecol.* 47: 416–432.

Büsgen, M., and E. Münch. 1929. *The Structure and Life of Forest Trees,* translated by T. Thomson. John Wiley & Sons, New York.

BIBLIOGRAPHY

Coile, T. S. 1940. Soil changes associated with loblolly pine succession on abandoned agricultural land of the Piedmont Plateau. *Duke Univ. School Forest. Bull.* 5: 1–85.

Coombe, D. E. 1957. The spectral composition of shade light in woodlands. *J. Ecol.* 45: 823–830.

Courant, R., and D. Hilbert. 1953. *Methods of Mathematical Physics. Volume I.* Interscience, New York.

Daubenmire, R. 1968. *Plant Communities: A Textbook of Plant Synecology.* Harper and Row, New York.

Decker, J. P. 1952. Tolerance is a good technical term. *J. Forestry* 50: 40–41.

Eliasson, L. 1968. Dependence of root growth on photosynthesis in *Populus tremula. Physiol. Plant.* 21: 806–810.

Evans, G. C. 1939. Ecological studies on the rain forest of southern Nigeria, II. The atmospheric environmental conditions. *J. Ecol.* 27: 436–482.

————. 1956. An area survey method of investigating the distribution of light intensity in woodlands with particular reference to sunflecks. *J. Ecol.* 44: 391–428.

Forbes, R. D., ed. 1955. *Forestry Handbook.* Ronald Press, New York.

Greenhill, A. G. 1881. Determination of the greatest height consistent with stability that a vertical pole or mast can be made, and the greatest height to which a tree of given proportions can grow. *Proc. Cambridge Phil. Soc.* 4: 65–73.

Grime, J. P., and D. W. Jeffrey. 1964. Seedling establishment in vertical gradients of sunlight. *J. Ecol.* 53: 621–642.

Grummer, G. 1961. The role of toxic substances in the interrelationships between higher plants. Pp. 219–228 in *Mechanisms in Biological Competition, S. E. B. Symposium 15.* Academic Press, New York.

Harper, J. L. 1969. The role of predation in vegetational diversity. Pp. 43–62 in *Diversity and Stability in Ecological Systems. Brookhaven Symposia in Biol.* No. 22. Brookhaven National Laboratory, Upton, New York.

Haxo, F., and L. R. Blinks. 1950. Photosynthetic action spectra of marine algae. *J. Gen. Physiol.* 33: 389–422.

Heath, O. V. S. 1969. *The Physiological Aspects of Photosynthesis.* Stanford Univ. Press, Stanford.

Hellmers, H. 1964. An evaluation of the photosynthetic efficiency of forests. *Quart. Rev. Biol.* 39: 249–257.

Holmes, R. W. 1957. Solar radiation, submarine daylight, and photosynthesis. Pp. 109–128 in J. W. Hedgpeth, ed., *Treatise on Marine Ecology and Paleoecology, Memoir 67 Vol.* 1. Geological Soc. of America, New York.

Hopkins, B. 1962. The measurement of available light by the use of *Chlorella. New Phytol.* 61: 221–223.

Jackson, L. W. R. 1967. Effect of shade on leaf structure of deciduous tree species. *Ecol.* 48: 498–499.

Jahnke, L. S., and D. B. Lawrence. 1965. Influence of photosynthetic crown structure on potential productivity of vegetation, based primarily on mathematical models. *Ecol.* 46: 319–326.

Jane, F. W. 1956. *The Structure of Wood.* Adam and Charles Black, London.

Janzen, D. H. 1970. Herbivores and the number of tree species in tropical forests. *Am. Naturalist* 104: 501–528.

Jones, E. W. 1945. The structure and reproduction of the virgin forest in the north temperate zone. *New Phytol.* 44: 130–148.

Kendall, M. G., and P. A. P. Moran. 1963. *Geometrical Probability.* Charles Griffen and Co., London.

Knoerr, K. R., and L. W. Gay. 1965. Tree leaf energy balance. *Ecol.* 46: 17–24.

Kramer, P. J., and W. S. Clark. 1947. A comparison of photosynthesis in individual pine needles and entire seedlings at various light intensities. *Plant Physiol.* 22: 51–57.

Kramer, P. J., and J. P. Decker. 1944. Relation between light intensity and rate of photosynthesis of loblolly pine and certain hardwoods. *Plant Physiol.* 19: 350–358.

BIBLIOGRAPHY

Kramer, P. J., and T. T. Kozlowski. 1960. *Physiology of Trees.* McGraw-Hill, New York.

Langford, A. N., and M. F. Buell. 1969. Integration, identity and stability in the plant association. *Adv. in Ecol. Res.* 6: 84–135.

Levins, R. 1968. *Evolution in Changing Environments: Some Theoretical Explorations, Monogr. Population Biol.* 2. Princeton Univ. Press, Princeton.

Lindeman, R. L. 1942. The trophic-dynamic aspect of ecology. *Ecol.* 23: 399–418.

Loomis, W. E. 1965. Absorption of radiant energy by leaves. *Ecol.* 46: 14–17.

Lundegårdh, H. 1966. Action spectra and the role of carotenoids in photosynthesis. *Physiol. Plant.* 19: 754–769.

Lyr, H., H. Polster, and H.-J. Fiedler. 1967. *Gehölzphysiologie.* G. Fischer, Jena.

MacArthur, R. H., and J. H. Connell. 1966. *The Biology of Populations.* John Wiley & Sons, New York.

MacArthur, R. H., and H. S. Horn. 1969. Foliage profile by vertical measurements. *Ecol.* 50: 802–804.

MacArthur, R. H., and E. O. Wilson. 1967. *The Theory of Island Biogeography, Monogr. Population Biol.* 1. Princeton Univ. Press, Princeton.

Madgwick, H. A. I., and G. L. Brumfield. 1969. The use of hemispherical photographs to assess light climate in the forest. *J. Ecol.* 57: 537–542.

Martin, N. D. 1959. An analysis of forest succession in Algonquin Park, Ontario. *Ecol. Monogr.* 29: 187–218.

Miller, P. C. 1969. Tests of solar radiation models in three forest canopies. *Ecol.* 50: 878–885.

Monsi, M., and T. Saeki. 1953. Über den Lichtfaktor in den Pflanzengesellschaften und seine Bedeutung für die Stoffproduktion. *Jap. J. Bot.* 14: 22–52. *Fide* Verhagen et al.

Monteith, J. L. 1963. Gas exchange in plant communities.

Pp. 95–112 in L. T. Evans, ed., *Environmental Control of Plant Growth*. Academic Press, New York.

————. 1965. Light distribution and photosynthesis in field crops. *Ann. Bot. N. S.* 29: 17–37.

Niering, W. A. 1953. The past and present vegetation of High Point State Park, New Jersey. *Ecol. Monogr.* 23: 127–148.

Odum, E. P. 1959. *Fundamentals of Ecology*, with H. T. Odum. Saunders, New York.

————. 1969. The strategy of ecosystem development. *Science* 164: 262–270.

Okali, D. U. U. 1966. A comparative study of the ecologically related tree species *Acer pseudoplatanus* and *Fraxinus excelsior*, I. The analysis of seedling distribution. *J. Ecol.* 54: 129–141.

Perry, T. O., H. E. Sellers, and C. O. Blanchard. 1969. Estimation of photosynthetically active radiation under a forest canopy with chlorophyll extracts and from basal area measurements. *Ecol.* 50: 39–44.

Ross, R. 1954. Ecological studies on the rain forest of southern Nigeria, III. Secondary succession in the Shasha Forest Reserve. *J. Ecol.* 42: 259–282.

Simpson, E. H. 1949. Measurement of diversity. *Nature* 163: 688.

Smith, J. M. 1968. *Mathematical Ideas in Biology*. Cambridge Univ. Press, Cambridge.

Sokal, R. R., and F. J. Rohlf. 1969. *Biometry: The Principles and Practice of Statistics in Biological Research*. W. H. Freeman & Co., San Francisco.

Verhagen, A. M. W., J. H. Wilson, and E. J. Britten. 1963. Plant production in relation to foliage illumination. *Ann. Bot. N. S.* 27: 627–640.

Vogel, S. 1968. "Sun leaves" and "shade leaves": differences in convective heat dissipation. *Ecol.* 49: 1203–1204.

Ward, R. T. 1956. The beech forests of Wisconsin—

changes in forest composition and the nature of the beech border. *Ecol.* 37: 407–419.

White, A., P. Handler, and E. Smith. 1968. *Principles of Biochemistry.* McGraw-Hill, New York.

Whittaker, R. H. 1966. Forest dimensions and production in the Great Smoky Mountains. *Ecol.* 47: 103–121.

———. 1969. Evolution of diversity in plant communities. Pp. 178–196 in *Diversity and Stability in Ecological Systems. Brookhaven Symposia in Biol.* No. 22. Brookhaven National Laboratory, Upton, New York.

Index

Abies, 2, 60f, 94f
Acer, 20–22, 24, 27, 29, 31–41, 61, 92, 100, 124, 126–28
adaptive strategies, 118ff
age distribution, 38, 85, 121, 125f
Anderson, M. C., 8, 15

Baker, F. S., 34, 49, 70
Bard, G. E., 22
Baumgartner, A., 57
Beer's law, 73
Betula, 3, 20–23, 27f, 31, 33f, 39, 61, 92f, 95, 106, 119, 124
Biswell, H. H., H. Buchanan, and R. P. Gibbens, 95
Blanchard, C. O., T. O. Perry, and H. E. Sellers, 14
Blinks, L. R., and F. Haxo, 11
Bormann, F. H., 61, 80
Boysen-Jensen, P., 69
branching pattern, 2, 90, 96, 101, 105
Britten, E. J., A. M. W. Verhagen, and J. H. Wilson, 79
Brockman, C. F., 131
Brumfield, G. L., and H. A. I. Madgwick, 15
Buchanan, H., R. P. Gibbens, and H. H. Biswell, 95
Buell, M. F., and A. N. Langford, 129
Buell, M. F., A. N. Langford, D. W. Davidson, and L. H. Ohmann, 22
Büsgen, M., and E. Münch, 49, 60, 81, 97, 101, 105

Calaveras County Big Tree Park (California), 94
canopy: effect on saplings, 15, 31; measurement of coverage, 11ff; preference of saplings, 32ff, 126f; shade cast, 30f, 56f, 84

Carya, 10f, 20–22, 26–28, 31, 33–36, 40, 91f, 101, 124, 127f
Cecropia, 96
Cercis, 127f
Chlorella, 14
chlorophyll, 14
chloroplasts, 81
Clark, W. S., and P. J. Kramer, 59, 69
clear-cut lumbering, 120
climax: cyclic, 35; definition, 3; diversity, 126; dominant species, 19, 85, 92f, 115; monolayered, 58; multilayered, 59, 93f, 115, 125; New Jersey, 22; productivity, 85, 93; stability, 85, 125; storied, 102; temperate, 61; xeric, 59
cloudy climate, 47, 59
Coile, T. S., 120
compensation point, 48, 67, 70ff
competition, 2, 11, 15, 128
Connell, J. H., and R. H. MacArthur, 51
Coombe, D. E., 9
corals, 129
Cornus, 10, 21, 26–28, 31, 34–36, 40, 91f, 94, 100f, 124, 127f
Courant, R., and D. Hilbert, 112
Croton, 96

data: age distribution, 39, 127; annual growth of saplings, 33; annual ring of trees, 124; canopy preference, 34, 36, 127; chloroplasts, 82; diversity of saplings, 40; effective leaf area, 92; germination inhibitor, 41; lobed leaves, 60; number of layers, 91f, 94f, 100; photosynthesis of leaves, 69; shade cast by canopy, 31; shade cast by tree, 92, 94, 98f; spectra of leaves, 10
Daubenmire, R., 38, 59

141

Davidson, D. W., L. H. Ohmann, M. F. Buell, and A. N. Langford, 22
Decker, J. P., 34
Decker, J. P., and P. J. Kramer, 59, 69
developmental pattern, 59, 97, 100f, 105, 116, 119
diversity: climax forest, 126; saplings, 39
drought resistance, 54f, 57

edaphic conditions, 2, 35, 57, 59, 93
Eliasson, L., 16
epiphytes, 96, 125
Euler's equation, 112
Evans, G. C., 9, 57

factor analysis, 4
Fagus, 3, 16, 20–22, 25, 27, 29, 31–40, 61, 69, 81f, 92, 100, 119, 124, 126–28
Finca Las Cruces (Costa Rica), 95
fish-eye photographs, 15, 28f
flat branches, 106
Forbes, R. D., 107
Fraxinus, 33f, 36, 90, 92, 100f, 126–28

Gay, L. W., and K. R. Knoerr, 55
germination inhibitors, 41
Giant Sequoia, 94
Gibbens, R. P., H. H. Biswell, and H. Buchanan, 95
Goreau, T. F., 129
Greenhill, A. G., 108
Grime, J. P., and D. W. Jeffrey, 59
growth: monolayer vs. multilayer, 55, 57, 75; saplings, 31, 101; trees, 57, 59, 84, 93, 101, 123
Grummer, G., 41

hardwood, 107, 113
Harper, J. L., 129
Haxo, F., and L. R. Blinks, 11
heartwood, 115
heat load, 4, 54f, 57, 79, 125
Heath, O. V. S., 68
Hellmers, H., 85

herbs, 96f, 121
Hilbert, D., and R. Courant, 112
Holmes, R. W., 11
Hopkins, B., 14
Horn, H. S., and R. H. MacArthur, 51

Institute Woods (New Jersey), 20ff
invasion, 84

Jackson, L. W. R., 81
Jahnke, L. S., and D. B. Lawrence, 108, 112
Jane, F. W., 107f
Janzen, D. H., 129
Jeffrey, D. W., and J. P. Grime, 59
Jones, E. W., 59
Juglans, 41
Juniperus, 3, 22, 60f, 119, 126f

Kendall, M. G., and P. A. P. Moran, 110
Knoerr, K. R., and L. W. Gay, 55
Kozlowski, T. T., and P. J. Kramer, 69, 97
Kramer, P. J., 59, 69, 97

Lactuca, 41, 43
Langford, A. N., and M. F. Buell, 129
Langford, A. N., D. W. Davidson, L. H. Ohmann, and M. F. Buell, 22
Larix, 61
Lawrence, D. B., and L. S. Jahnke, 108, 112
leaf size: climatic gradient, 47, 59; in multilayer, 47, 54, 90, 123; optimal, 54f, 57; shadow, 46; successional pattern, 57, 61; tropical, 96, 121
leaves: effective area index, 93, 123; horizontal distribution, 2, 51, 53f, 57, 64f; lobed, 54, 59f; optimal density, 55, 70, 72, 74, 76, 84; orientation, 2, 60f, 79; shadow, 45; shape, 54, 59f; sun and shade, 59, 80; transmission spectra, 9; vertical distribution, 64, 70f, 79. See also leaf size

Levins, R., 119
Libocedrus, 94f
Light: integrator, 14; meter, 12; spectral quality, 9
Lindeman, R. L., 122
Liquidambar, 21f, 31, 33f, 36, 39–41, 61, 124
Liriodendron, 21, 31, 39f, 92, 124, 127f
Loomis, W. E., 10, 48
Lundegårdh, H., 11
Lyr, H., H. Polster, and H.-J. Fiedler, 61

MacArthur, R. H., and J. H. Connell, 51
MacArthur, R. H., and H. S. Horn, 51
MacArthur, R. H., and E. O. Wilson, 128
Madgwick, H. A. I., and G. L. Brumfield, 15
Mammoth Cave National Park (Kentucky), 34, 126
Martin, N. D., 59
measurement of: layers, 89ff; light intensity, 8ff; projection, 8ff; shade cast, 11ff; successional status, 37ff; tolerance, 34
Michaelis-Menten equation, 66
Miller, P. C., 73
monolayer: definition, 53, 65; tropical, 96, 121; vs. multilayer, 53ff, 74ff
Monsi, M., and T. Saeki, 73
Monteith, J. L., 68, 73
Moran, P. A. P., and M. G. Kendall, 110
multilayer: definition, 53, 65; ideal, 79, 123; persistent, 59, 93f, 115, 125; vs. monolayer, 53ff, 74ff
Münch, E., and M. Büsgen, 49, 60, 81, 97, 101, 105
Myrtus, 2

Niering, W. A., 22
nutrients, 4, 123
Nyssa, 21f, 31, 33–40, 61, 92, 100f, 119, 124, 127f

Ochroma, 96
Odum, E. P., 122f
Ohmann, L. H., M. F. Buell, A. N. Langford, and D. W. Davidson, 22
Okali, D. U. U., 101
old-field succession, 97
Ostrya, 126–28
overmaturity, 85

palms, 105
Perry, T. O., H. E. Sellers, and C. O. Blanchard, 14
photography, 12, 15
photosynthesis, 48ff, 64ff; implications of saturation, 48; temporal course, 80
phytoplankton, 123
Picea, 3, 21, 61
Pinus, 3, 22, 48, 59, 61, 90, 92–96, 98f, 102, 120
Piper, 96
pith, 115
Podophyllum, 10f, 27, 91
Poisson distribution, 51
Populus, 21, 31, 33–35, 39f, 61, 90, 92f, 95, 106, 124
predation, 125, 129
productivity, 85, 93, 122f
projection: branches, 89; canopy, 28f, 31; leaves, 51ff, 64f; measurement, 9ff; tree, 57, 65, 89, 108ff

Quercus, 3, 10f, 20–22, 26–28, 31, 34–36, 39–41, 48, 59–61, 69f, 91f, 101, 124, 127f

random distribution, 51, 65
Rohlf, F. J., and R. R. Sokal, 30, 36
Roos, P. J., 129
root competition, 11, 15, 86, 119
Ross, R., 58

Saeki, T., and M. Monsi, 73
sapwood, 115
Salix, 90
Sassafras, 31, 33f, 36, 40f, 43, 91f, 100, 121, 124, 127f
self-pruning, 97

Sellers, H. E., C. O. Blanchard, and T. O. Perry, 14
Sequoia, 60, 94, 102
shade leaves, 59, 80
shapes of trees, 104ff
Simpson, E. H., 39
Smith, J. M., 53
snow, 106
softwood, 107, 113
Sokal, R. R., and F. J. Rohlf, 30, 36
sprouts: epicormic, 97; root, 37, 85; stump, 120
stability, 38, 85, 121, 125f
storied forest, 102
storms, 125
succession: Costa Rica, 95f; definition, 3; Kentucky, 126–28; multilayer to monolayer, 58ff, 92, 95; New Jersey, 22, 37ff; old-field, 97; patterns in, 122ff; primary, 59; simulation, 34f; speed of changes, 75; temperate, 61
sun leaves, 59, 80
sunburn, 11, 105
surface area of tree crown, 112

Taxus, 2
Taylor series, 52
Theophrastus of Eresos, 2
tolerance, 2, 19, 34, 56–58
tree-ferns, 105
tropical vegetation, 95f, 120f
Tsuga, 61, 92, 96, 98–100

Ulmus, 90
understory, 11, 27, 57, 91, 102, 125

Verhagen, A. M. W., J. H. Wilson, and E. J. Britten, 79
Viburnum, 10, 27, 91
vines, 59, 96, 125
virgin forest, 85, 93
Vogel, S., 55, 59

Ward, R. T., 37
water balance, 4
White, A., P. Handler, and E. Smith, 66
White Pine, 93, 120
Whittaker, R. H., 85, 123, 129
Wiesner, 97
Wilson, E. O., and R. H. MacArthur, 128
Wilson, J. H., E. J. Britten, and A. M. W. Verhagen, 79
wind, 2, 55, 106, 108
windblown seeds, 85, 119
wood, 107f, 113ff

xeric environment, 35, 57, 59, 93, 95

yew, 2

zoophytes, 129

Milton Keynes UK
Ingram Content Group UK Ltd.
UKHW010153100124
435723UK00001B/5

9 780691 023557